Atlas of the Human Eye

Atlas of the Human Eye

Anatomy & Biometrics

Héctor Barajas M.

Library of Congress Control Number:		2015920179
ISBN:	Hardcover	978-1-5065-1034-7
	Softcover	978-1-5065-1033-0
	eBook	978-1-5065-1032-3

Print information available on the last page.

Rev. date: 14/01/2016

To order additional copies of this book, please contact:

Palibrio

1663 Liberty Drive

Suite 200

Bloomington, IN 47403

Toll Free from the U.S.A 877.407.5847

Toll Free from Mexico 01.800.288.2243

Toll Free from Spain 900.866.949

From other International locations +1.812.671.9757

Fax: 01.812.355.1576

orders@palibrio.com

729814

CONTENTS

PREFACE

The purpose of this work is to present to professionals of the visual sciences the results obtained in the research of anatomical dimensions of the multiple structures of the eye obtained in vivo. This research was completed through utilizing current auxiliary instruments in ophthalmological diagnosis that utilized microns as units and thus, allowed obtaining high optical resolution images.

The figures obtain through this research may not coincide with the traditional concepts. It is necessary to take into consideration factors that may cause tissue modifications such as race, age, physiological variants and changes caused by various pathologies. It is important to note that in addition to image, a thorough investigation with a complete clinical history is required in order to obtain a more conclusive diagnosis

New technological tools that are constantly improving accuracy have allowed us to discover tissues and structures, explore new spaces, define measurements and evolve concepts.

HB

E-mail: medicusocularis@hotmail.com

ACKNOWLEDGMENTS

Gratitude to Dr. Charles J. Pavlin (1944-2014) for it was his enthusiasm and made invaluable comments and suggestions.

REVIEW BOARD

GENERAL ANATOMY

Eyeball

1.- Fibrous tunic
1.1.- Cornea
a) epithelium
b) Bowman's layer
c) stroma
d) Descemet's membrane
e) endothelium
1.2.- Sclera
a) episclera
b) stroma
c) lamina fusca

2.- Vascular tunic
2.1.- Iris
2.2.- Ciliary body
2.2.1.- anterior pars plicata
2.2.1.1.- ciliary muscle
a) longitudinal fibres
b) radial fibres
c) circular fibres
2.2.1.2.- ciliary processes
2.2.2.- posterior pars plana
a) longitudinal fibres
2.3.- Choroid
a) suprachoroid
b) stroma
c) choriocapillaris
d) Bruch's membrane

3.- Nervous tunic-Retina
4.- Optic nerve
5.- Aqueous humor
6.- Lens
7.- Zinn's zonule
8.- Vitreous body

GENERAL ANATOMY

Eyeball

1.- Fibrous tunic
  1.1- cornea
    a) epithelium
    b) Bowman's layer
    c) stroma
    d) Descemet's membrane
    e) endothelium
  1.2.- sclera
    a) episclera
    b) stroma
    c) lamina fusca

2.- Vascular tunic
  2.1.- iris
  2.2.- ciliary body
    2.2.1.- anterior pars plicata
      2.2.1.1.- ciliary muscle
        a) longitudinal fibres
        b) radial fibres
        c) circular fibres
      2.2.1.2.- ciliary processes
    2.2.2.- posterior pars plana
      a) longitudinal fibres
  2.3.- choroid
    a) suprachoroid
    b) stroma
    c) choriocapillaris
    d) Bruch's membrane

3.- Nervous tunic-Retina
4.- Optic nerve
5.- Aqueous humor
6.- Lens
7.- Zinn´s zonule
8.- Vitreous body

ANATOMY

Eyeball

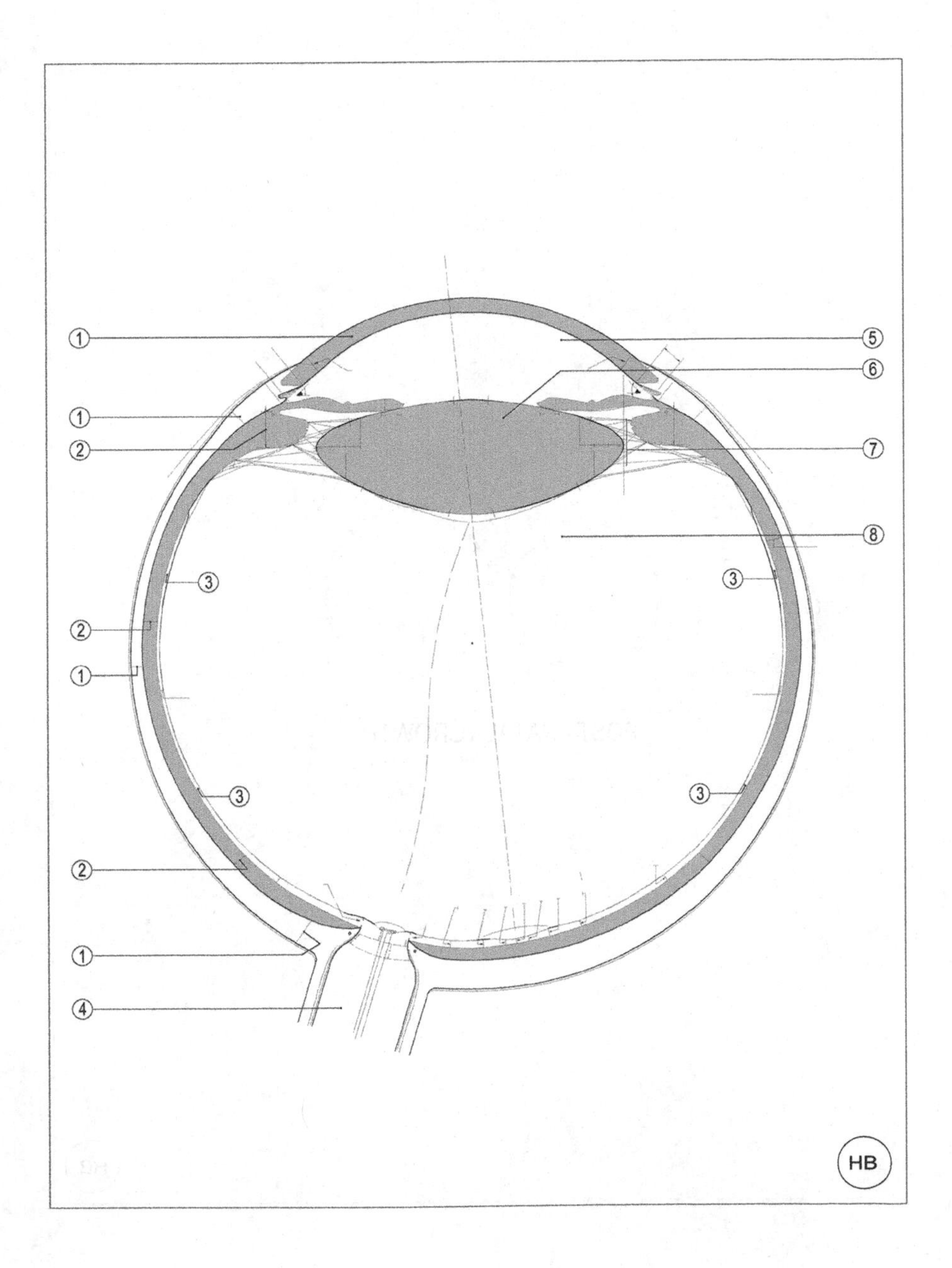

BIOMETRICS

Eyeball

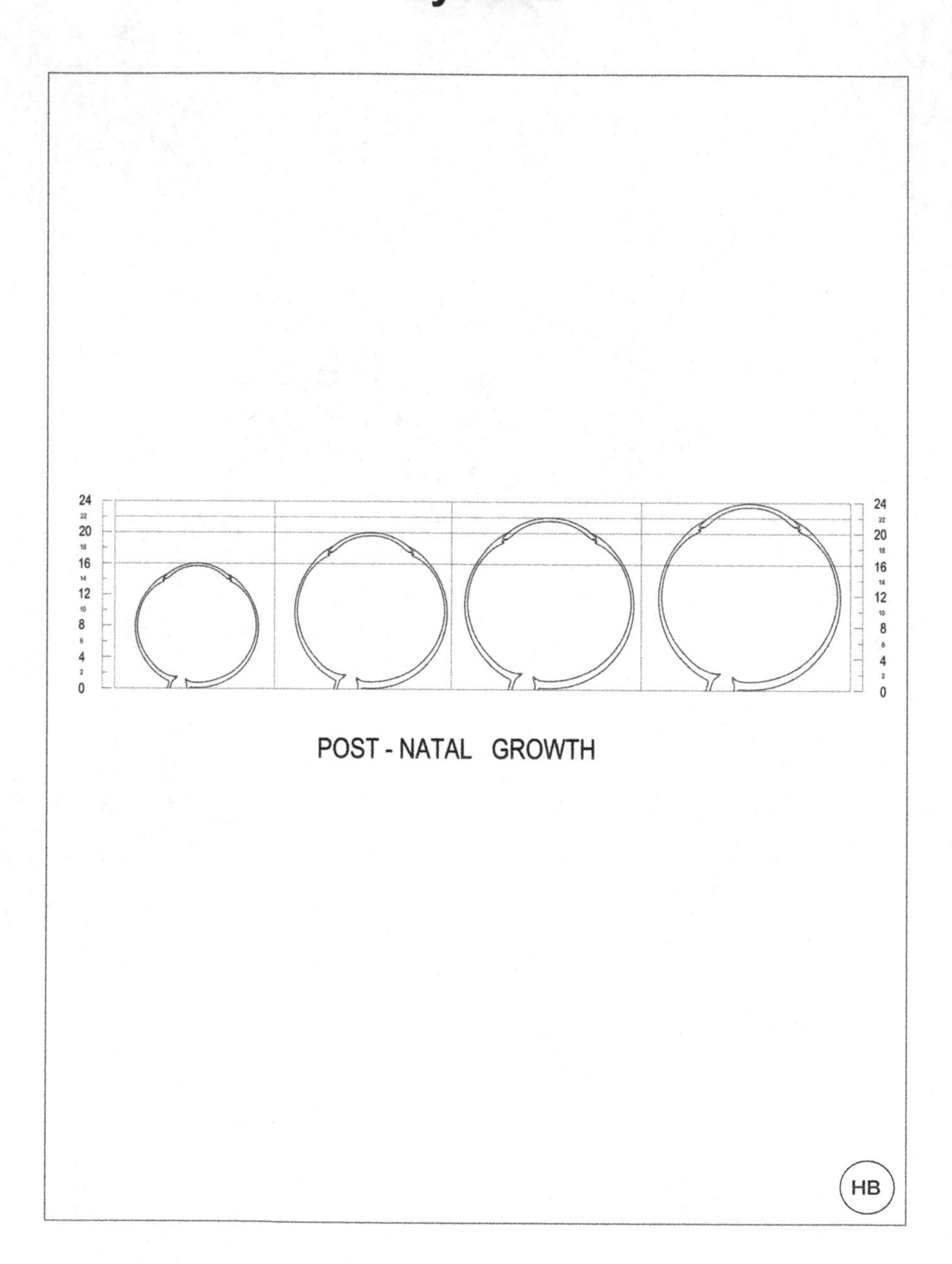

POST - NATAL GROWTH

1

BIOMETRICS

Eyeball

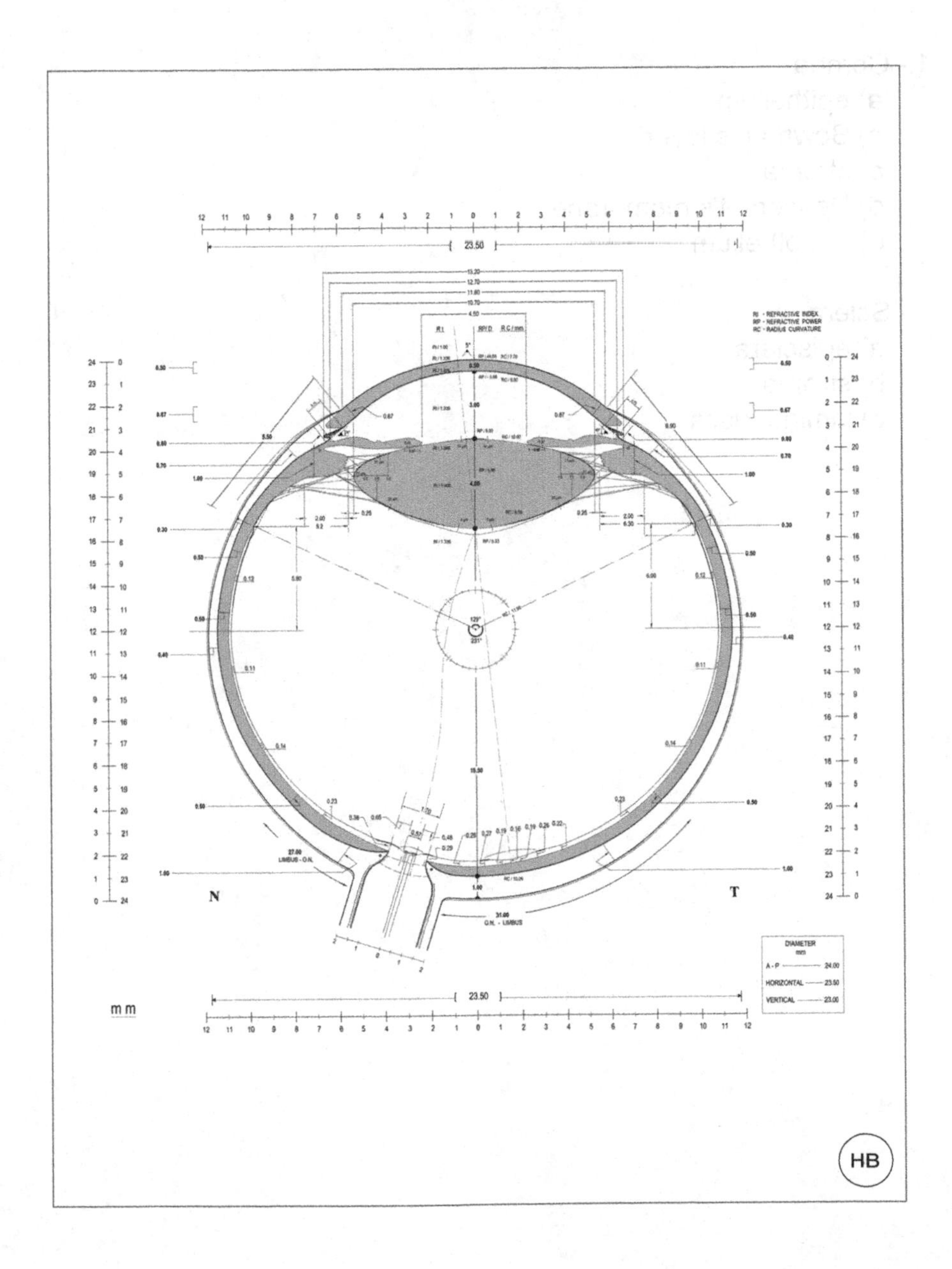

ANATOMY

Fibrous tunic

1.- Cornea
   a) epithelium
   b) Bowman's layer
   c) stroma
   d) Descemet's membrane
   e) endothelium

2.- Sclera
   a) episclera
   b) stroma
   c) lamina fusca

ANATOMY

Fibrous tunic

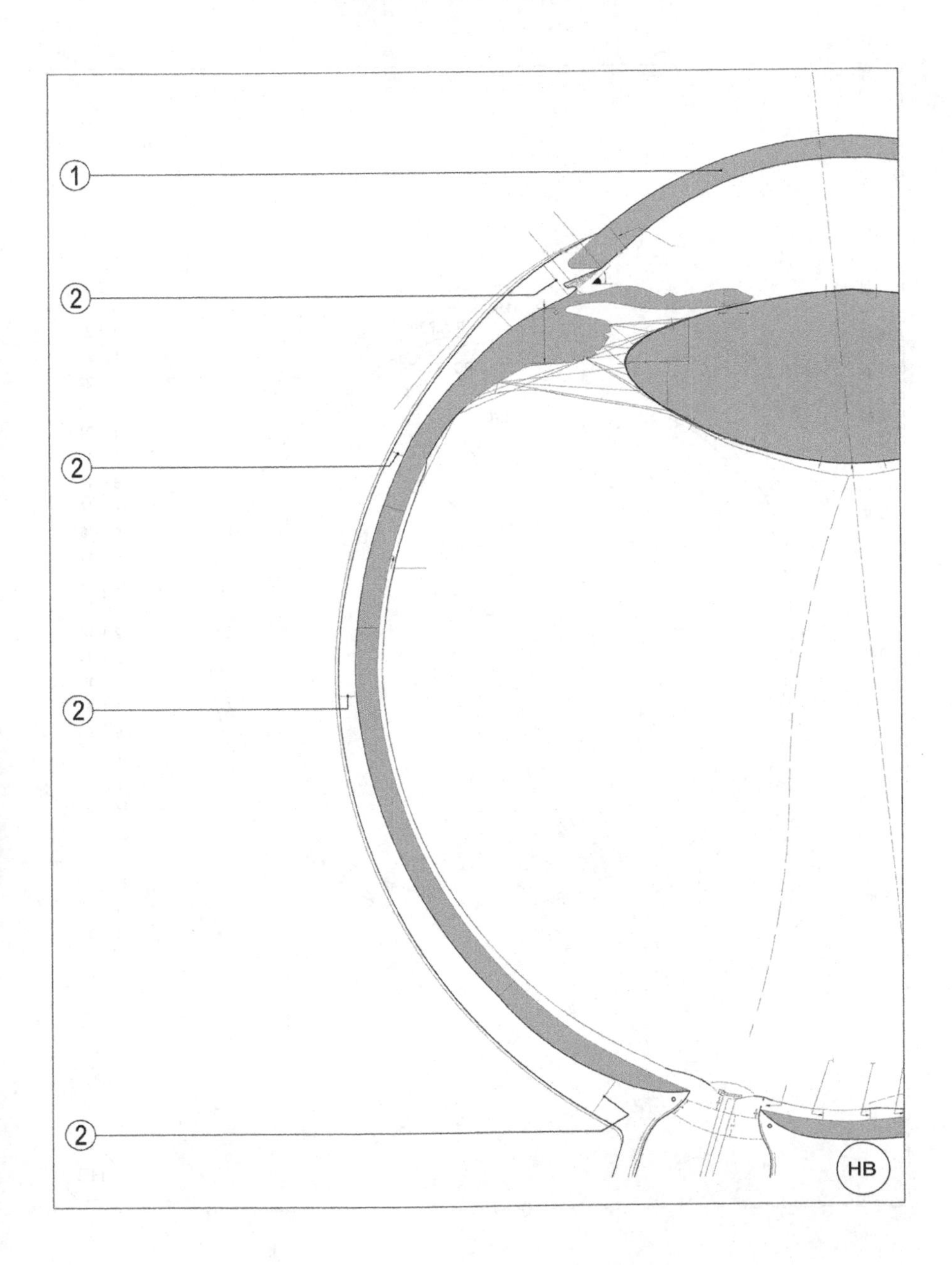

BIOMETRICS

Fibrous tunic

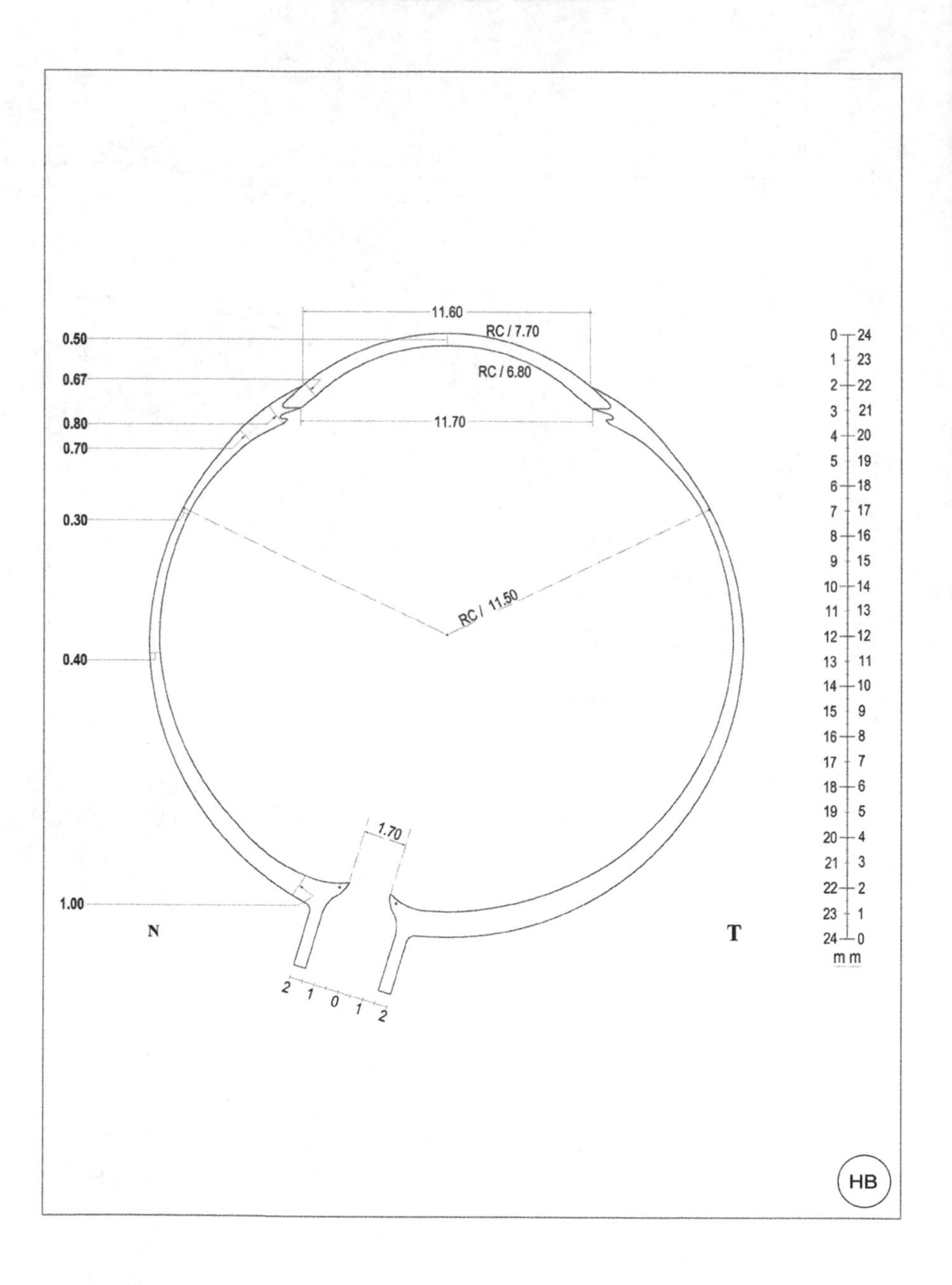

BIOMETRICS

Fibrous tunic

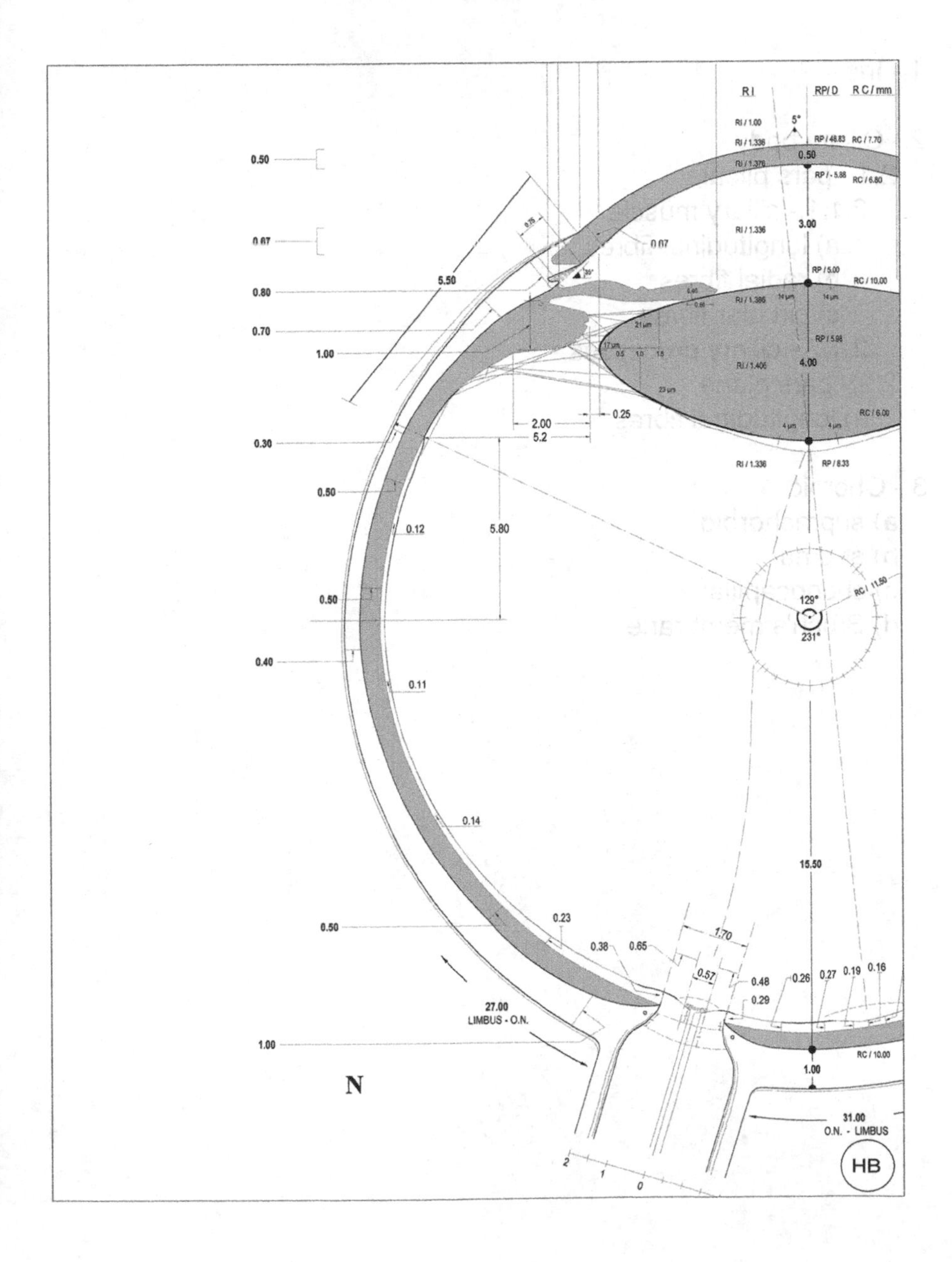

ANATOMY

Vascular tunic

1.- Iris

2.- Ciliary body
  2.1.- pars plicata
    2.1.1.- ciliary muscle
      a) longitudinal fibres
      b) radial fibres,
      c) circular fibres
    2.1.2.- ciliary processes.
  2.2.- pars plana
    a) longitudinal fibres

3.- Choroid
  a) suprachoroid
  b) stroma,
  c) choriocapillaris
  d) Bruch's membrane

ANATOMY

Vascular tunic

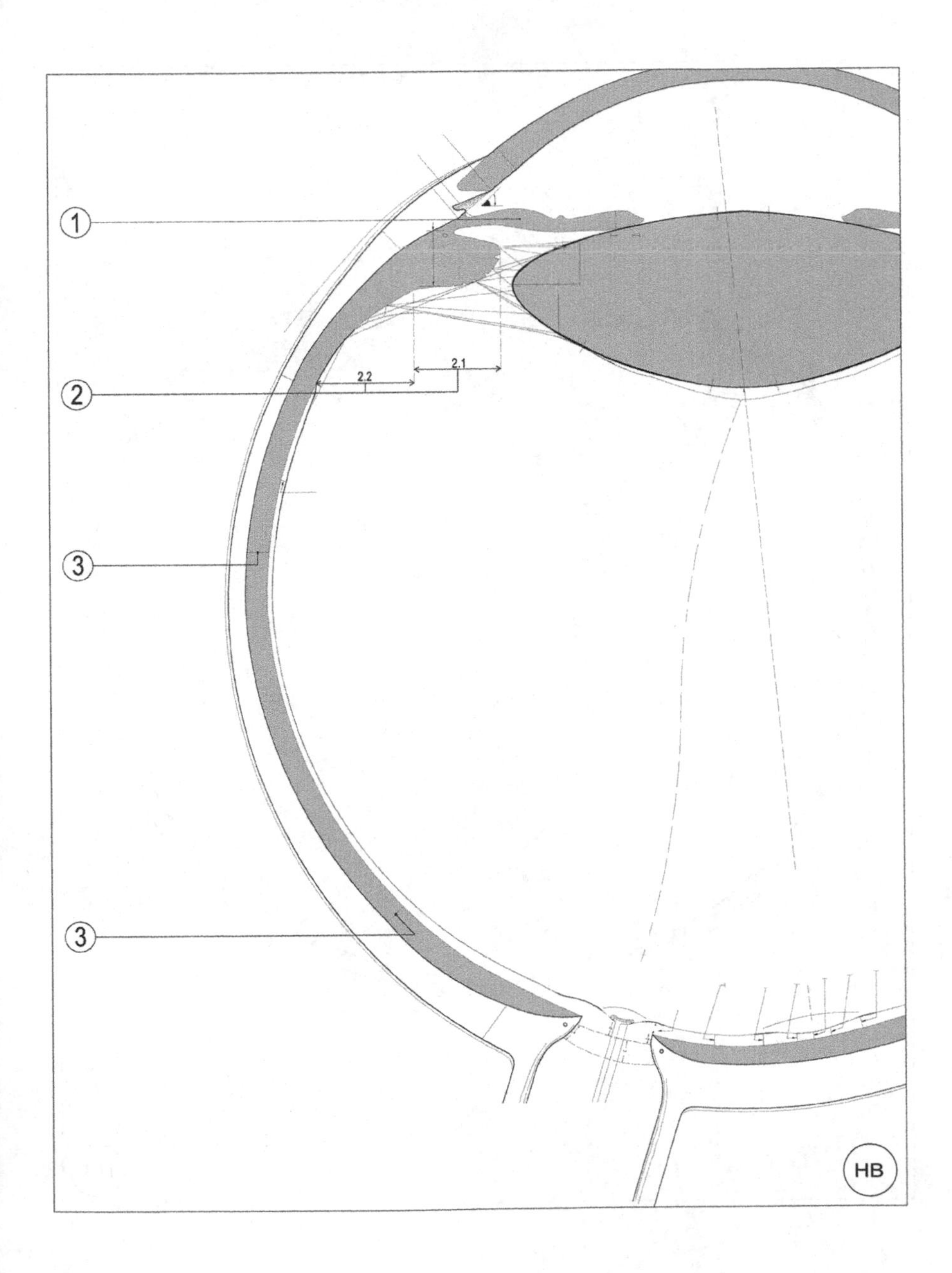

BIOMETRICS

Vascular tunic

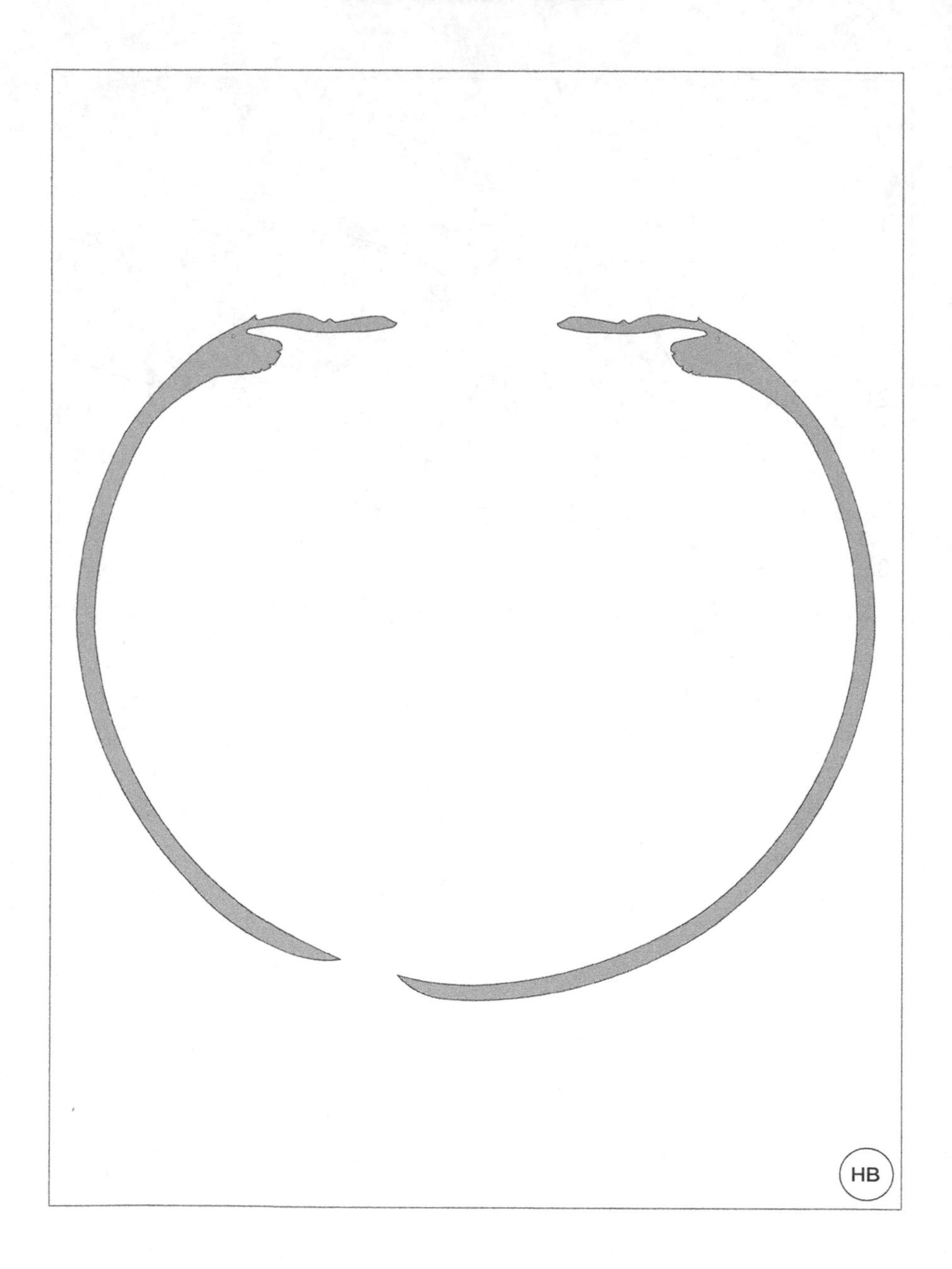

BIOMETRICS

Vascular tunic

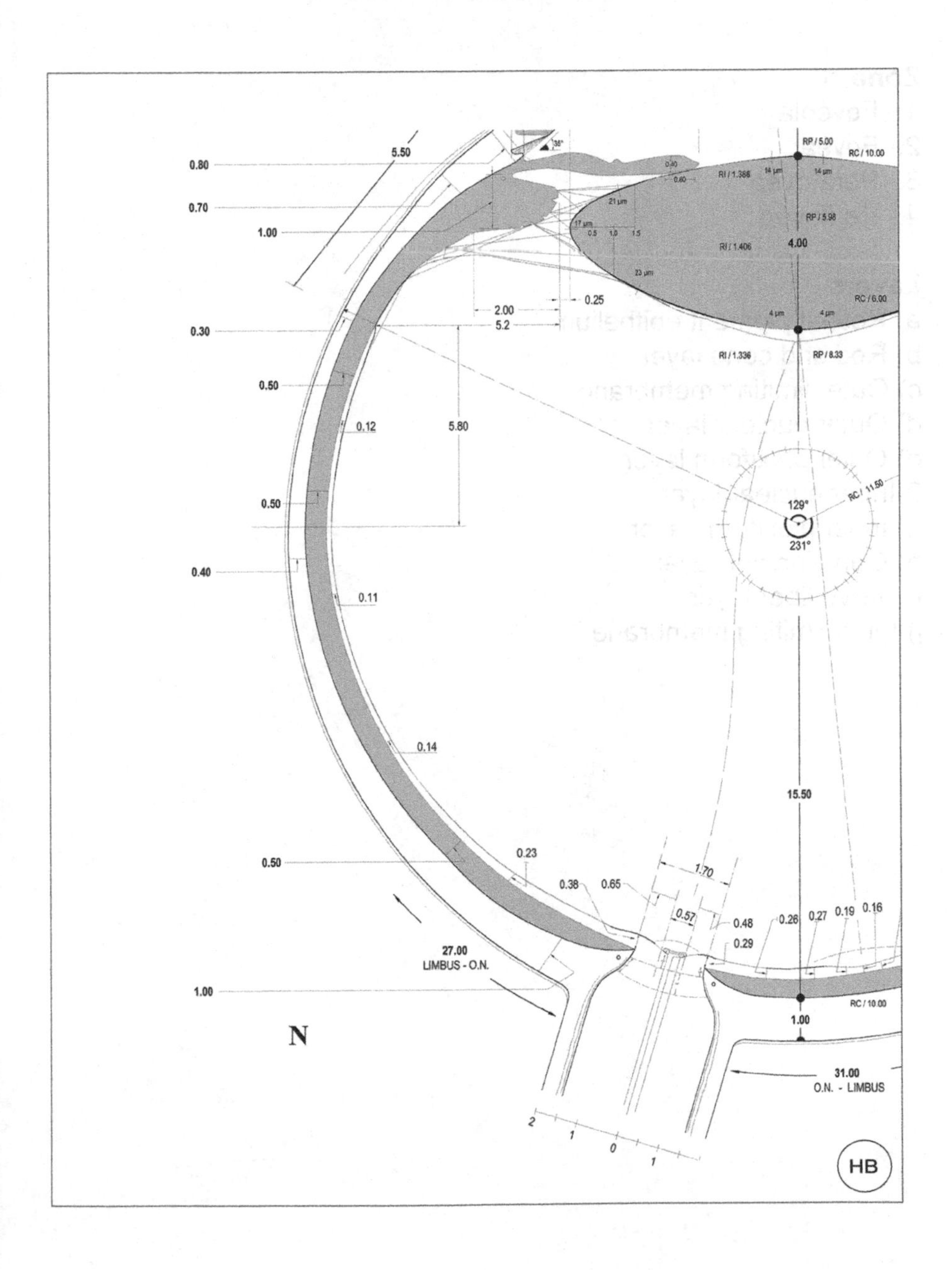

ANATOMY

Nervious tunic - Retina

Zone
1.- Foveola
2.- Fovea
3.- Parafovea
4.- Perifovea

Layers
a) Retinal pigment epithelium
b) Rod and cone layer
c) Outer limiting membrane
d) Outer nuclear layer
e) Outer plexiform layer
f) Inner nuclear layer
g) Inner plexiform layer
h) Ganglion cell layer
i) Nerve fiber layer
j) Inner limiting membrane

ANATOMY

Nervious tunic - Retina

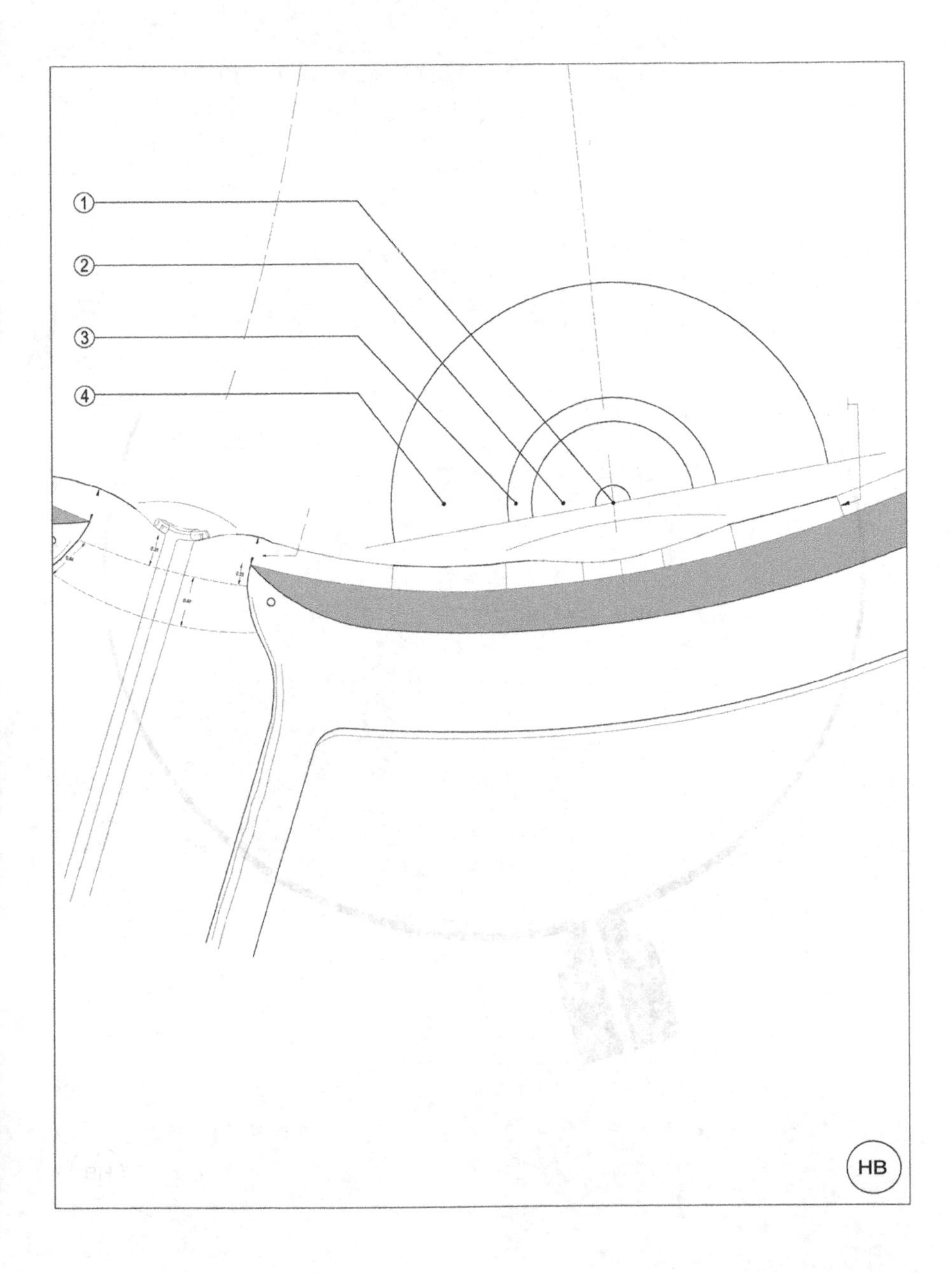

BIOMETRICS

Nervious tunic - Retina

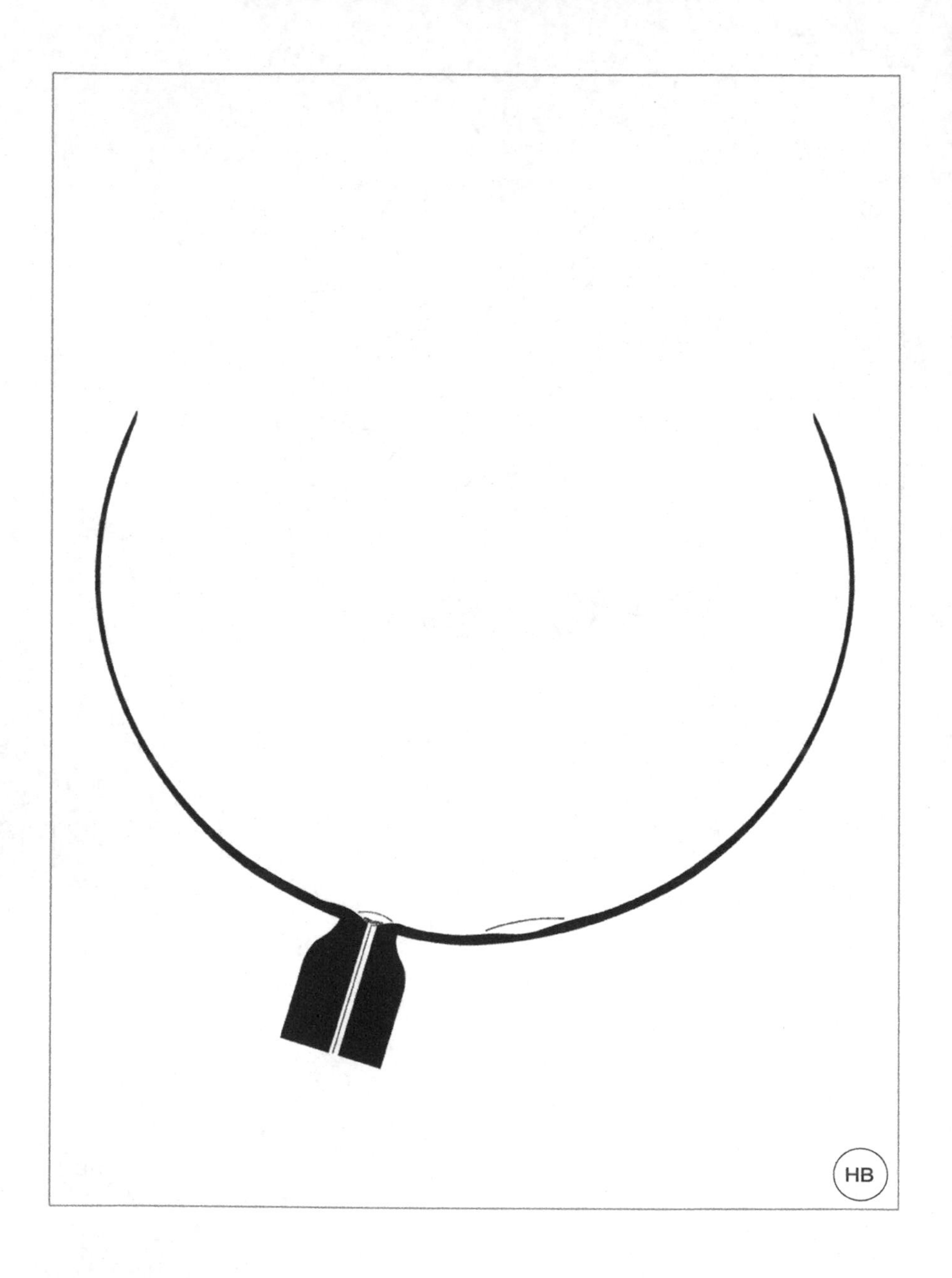

BIOMETRICS

Nervious tunic - Retina

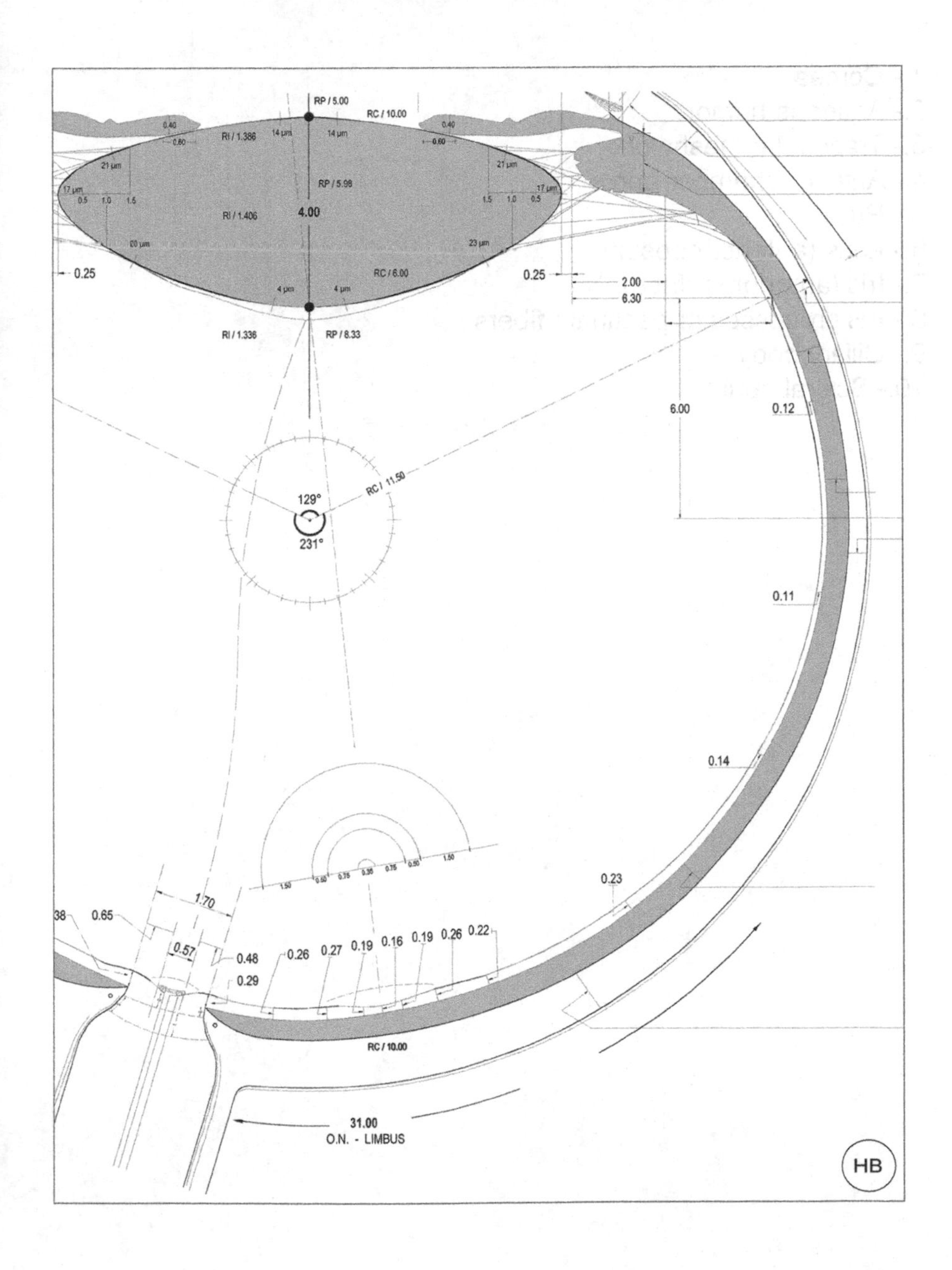

ANATOMY

Anterior chamber

1.- Cornea
2.- Aqueous humor
3.- Trabecular meshwork
4.- Anterior chamber angle
5.- Pupil
6.- Lens (anterior capsule)
7.- Iris (anterior surface)
8.- Iris processes or pectinate fibers
9.- Ciliary body
10.- Scleral spur

ANATOMY

Anterior chamber

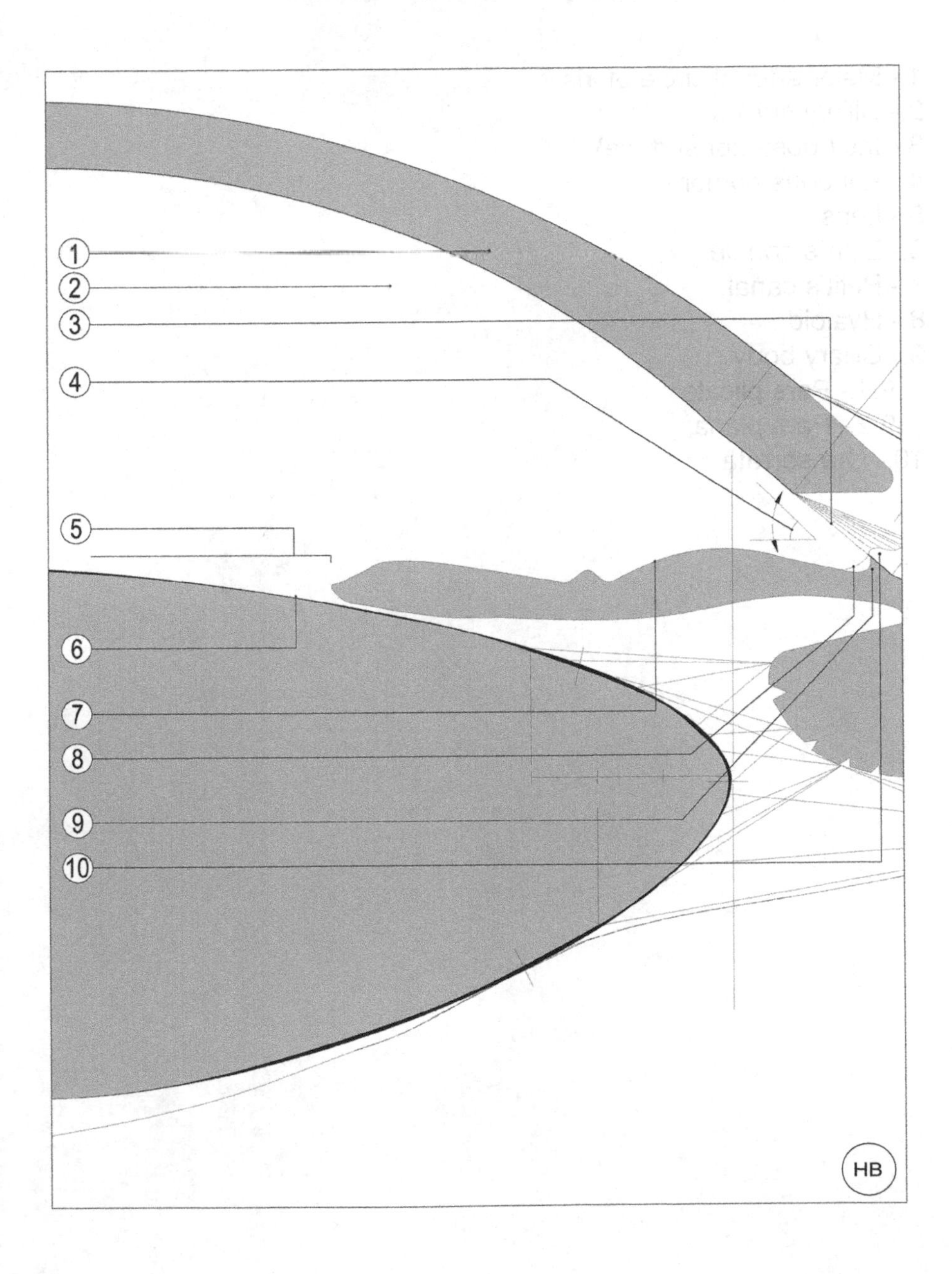

ANATOMY

Posterior chamber

1.- Major arterial circle of iris
2.- Ciliary sulcus
3.- Iris (posterior surface)
4.- Aqueous humor
5.- Lens
6.- Zinn´s zonule
7.- Petit's canal
8.- Hyaloid
9.- Ciliary body
   9.1.- Pars plicata
   9.2.- Pars plana
10.- Ora serrata

ANATOMY

Posterior chamber

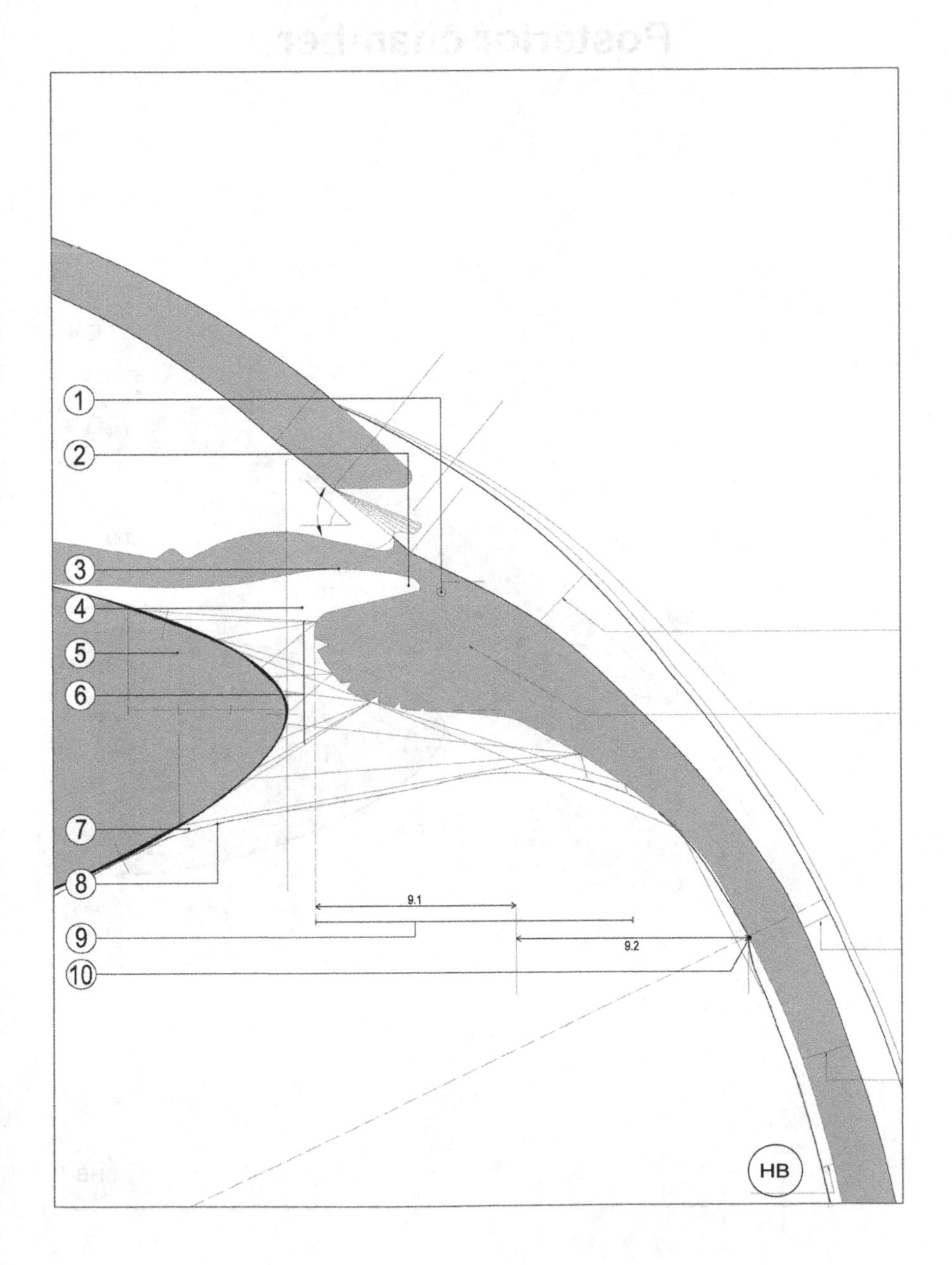

BIOMETRICS

Anterior chamber

Posterior chamber

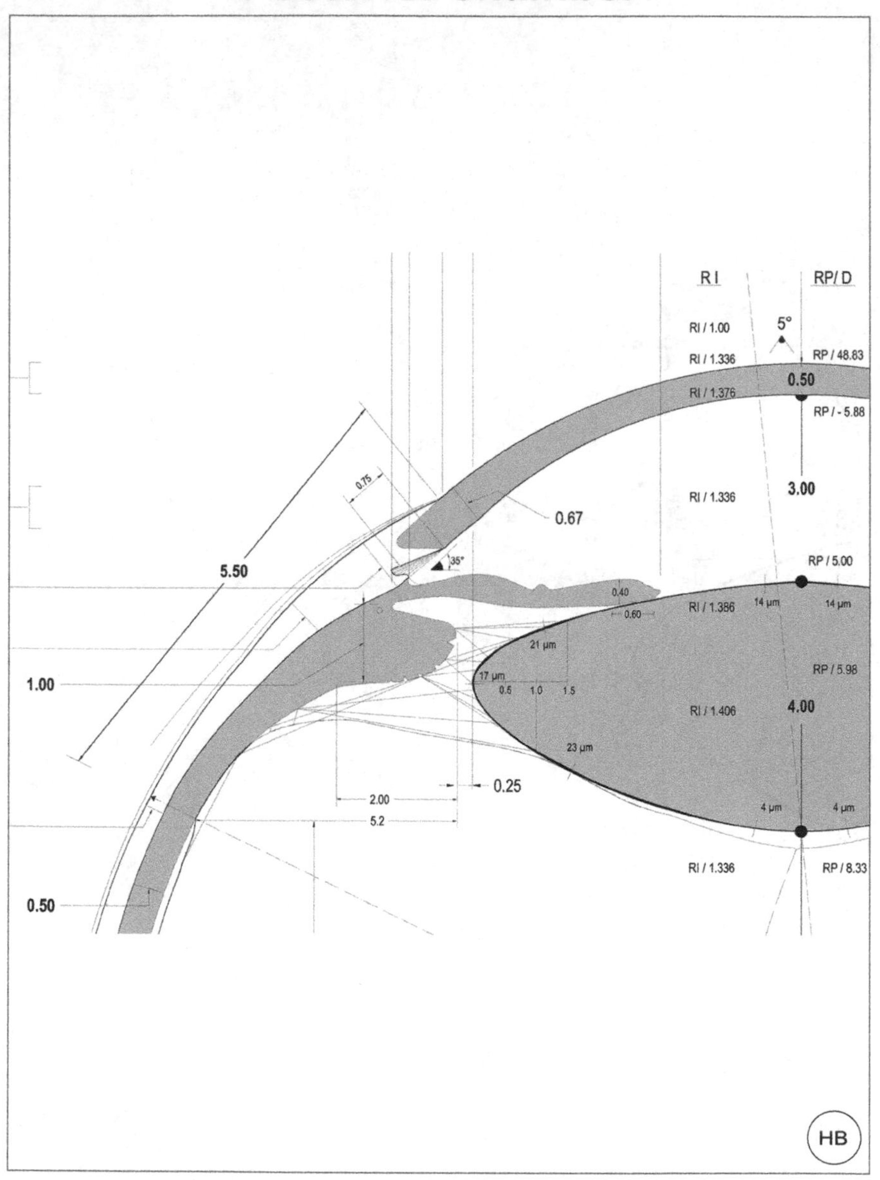

BIOMETRICS

Anterior chamber
Posterior chamber

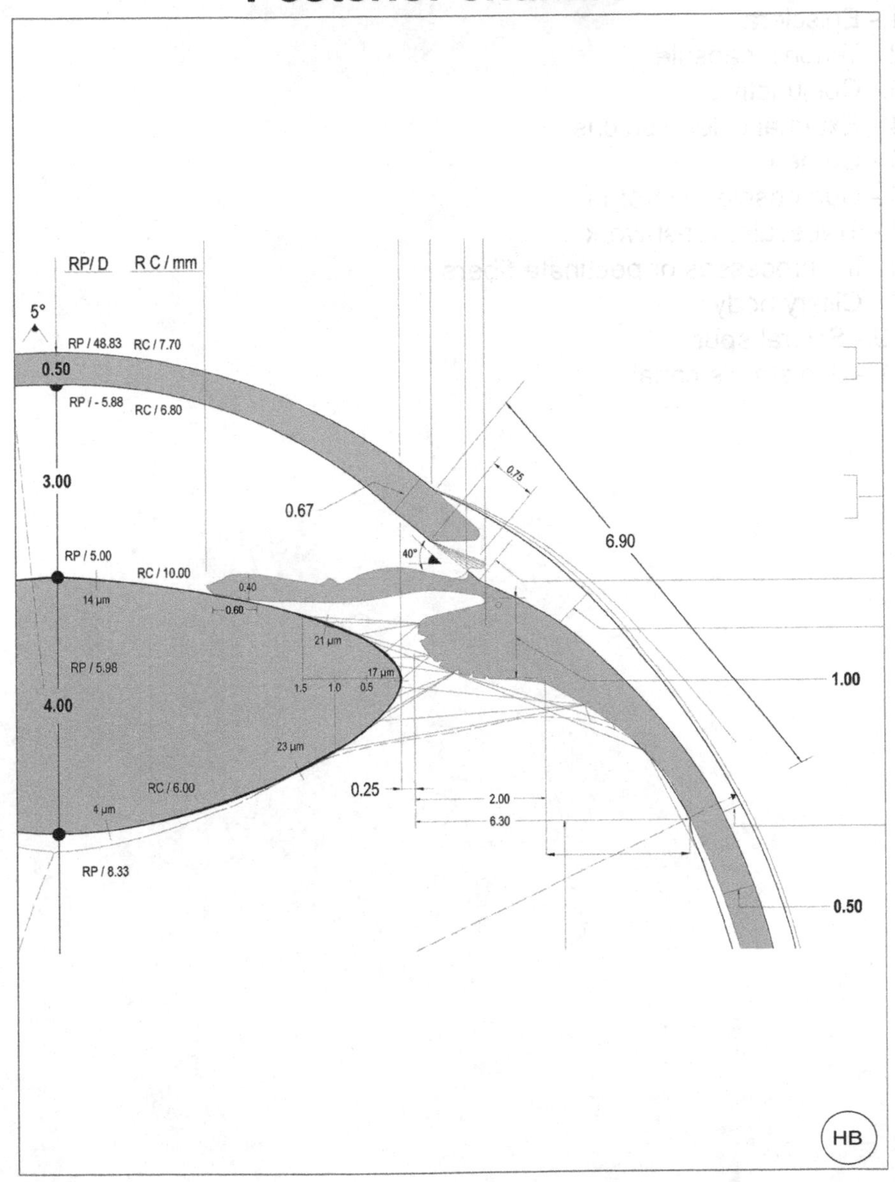

ANATOMY

Limbus

1.- Episclera
2.- Tenon's capsule
3.- Conjunctiva
4.- External scleral sulcus
5.- Cornea
6.- Corneoscleral stroma
7.- Trabecular meshwork
8.- Iris processes or pectinate fibers
9.- Ciliary body
10.- Scleral spur
11.- Schlemm's canal

ANATOMY

Limbus

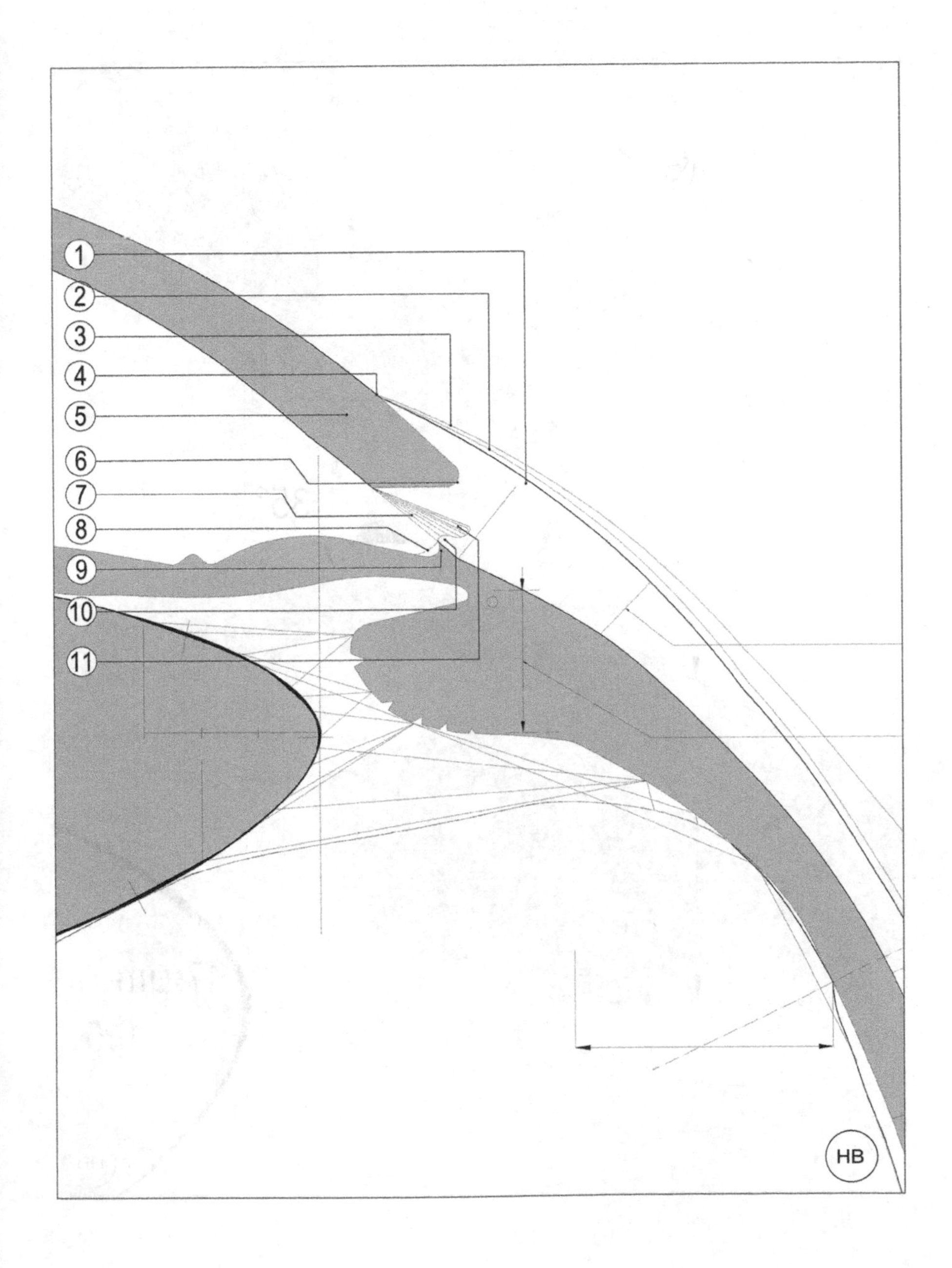

BIOMETRICS

Limbus

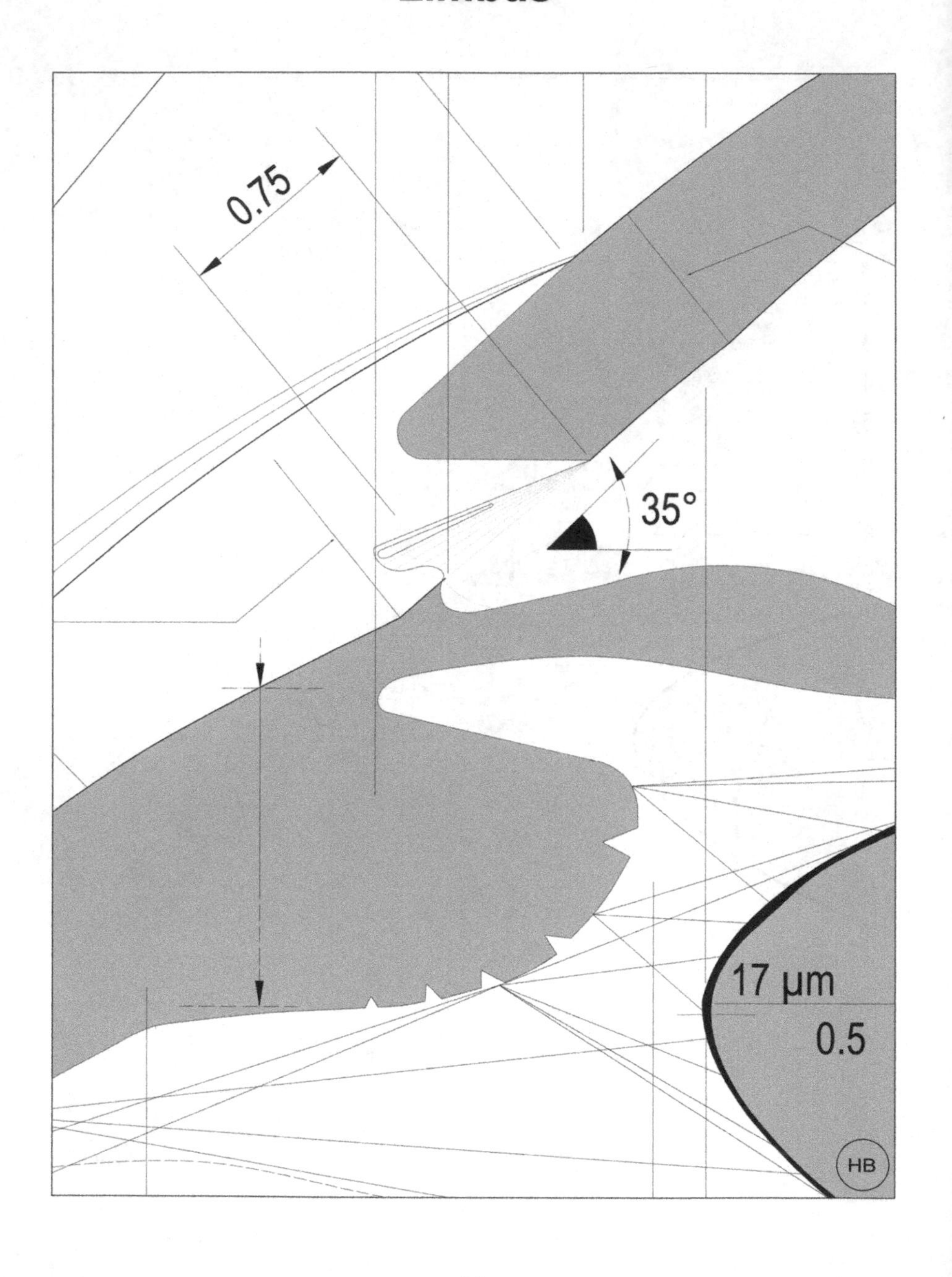

BIOMETRICS

Limbus

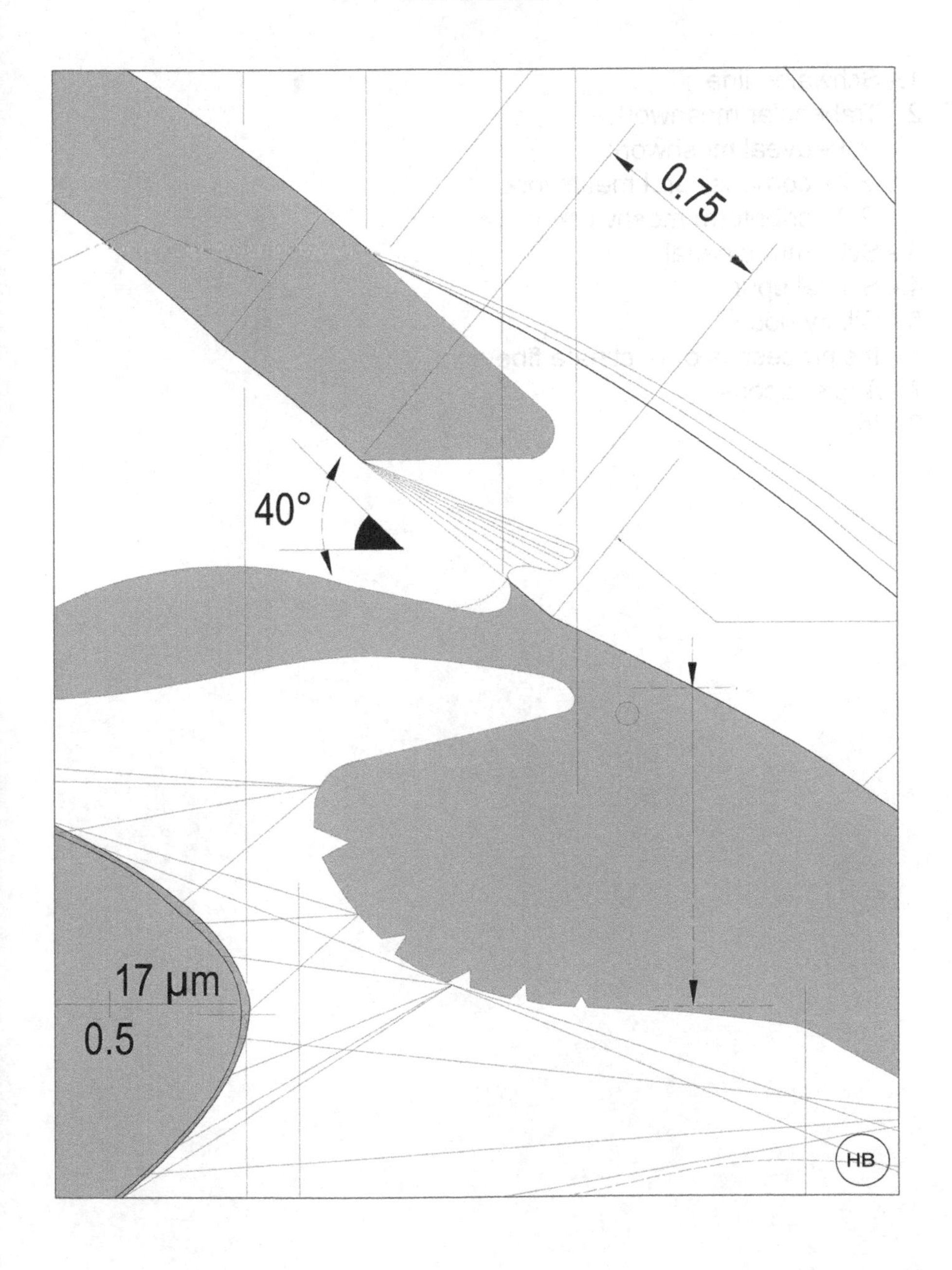

ANATOMY

Iridocorneal angle

1.- Schwalbe line
2.- Trabecular meshwork
  2.1.- uveal meshwork
  2.2.- corneoscleral meshwork
  2.3.- cribriform meshwork
3.- Schlemm´s canal
4.- Scleral spur
5.- Ciliary body
6.- Iris processes or pectinate fibers
7.- Angle recess
8.- Iris

ANATOMY

Iridocorneal angle

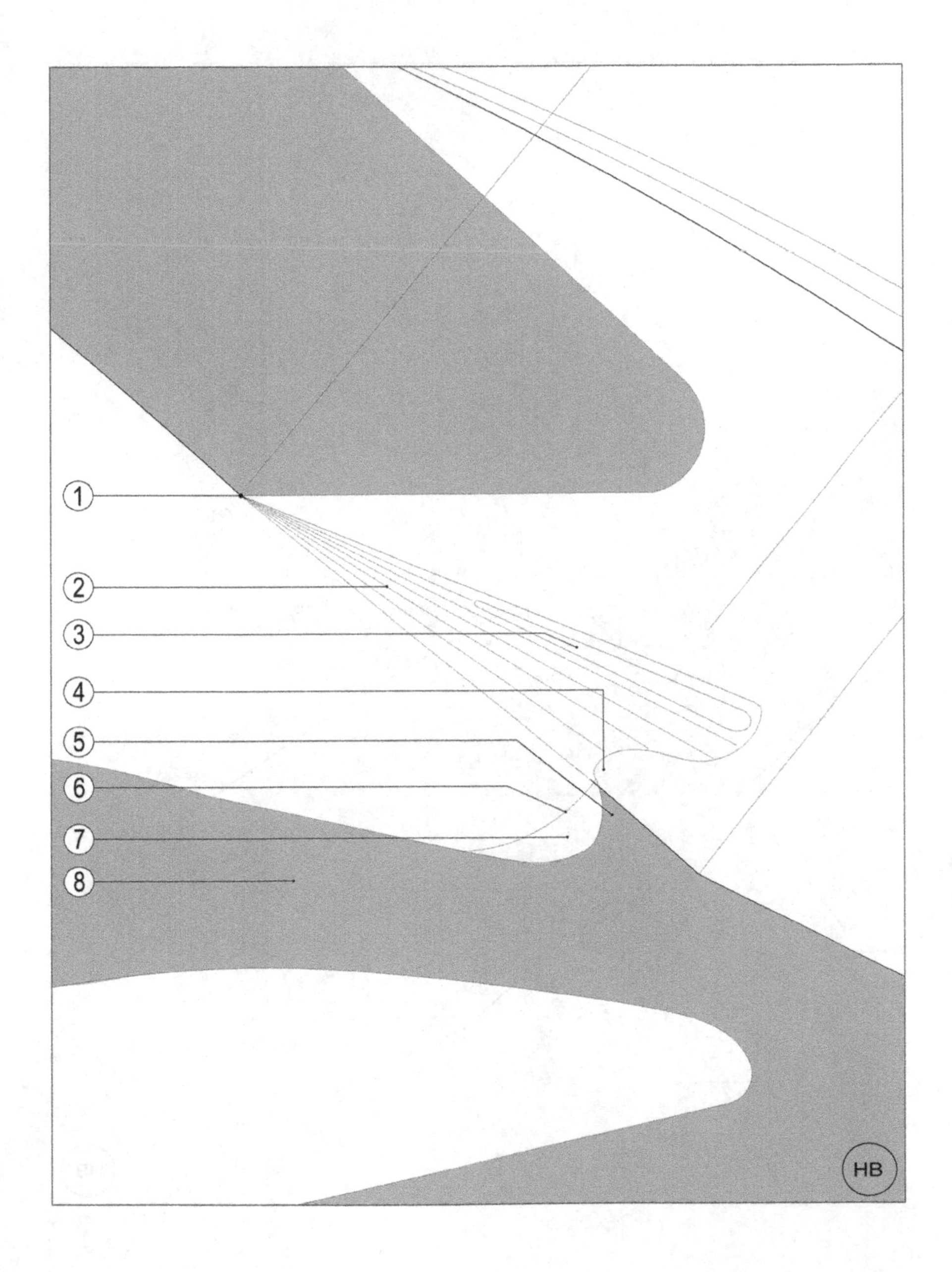

BIOMETRICS

Iridocorneal angle

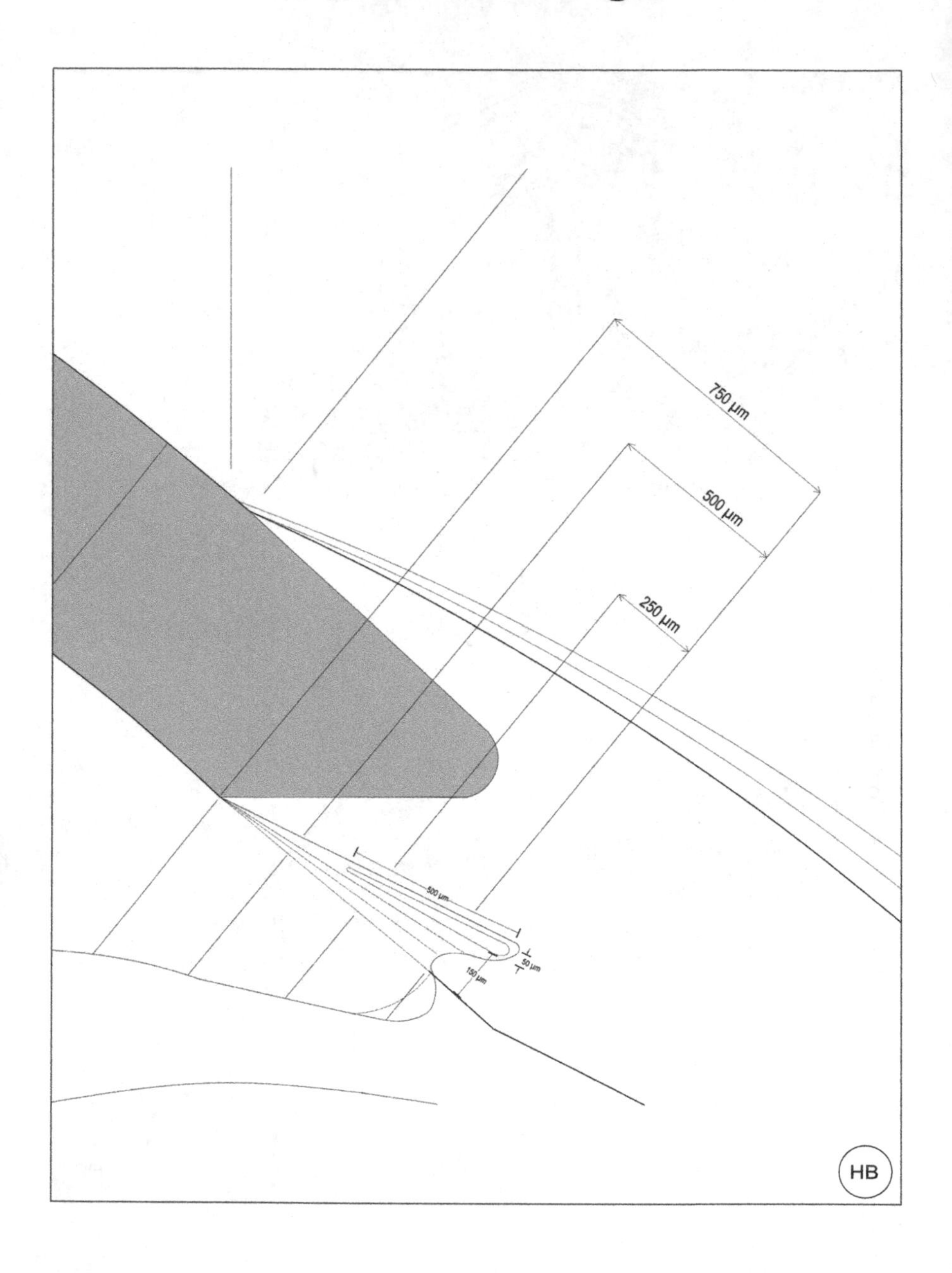

BIOMETRICS

Iridocorneal angle

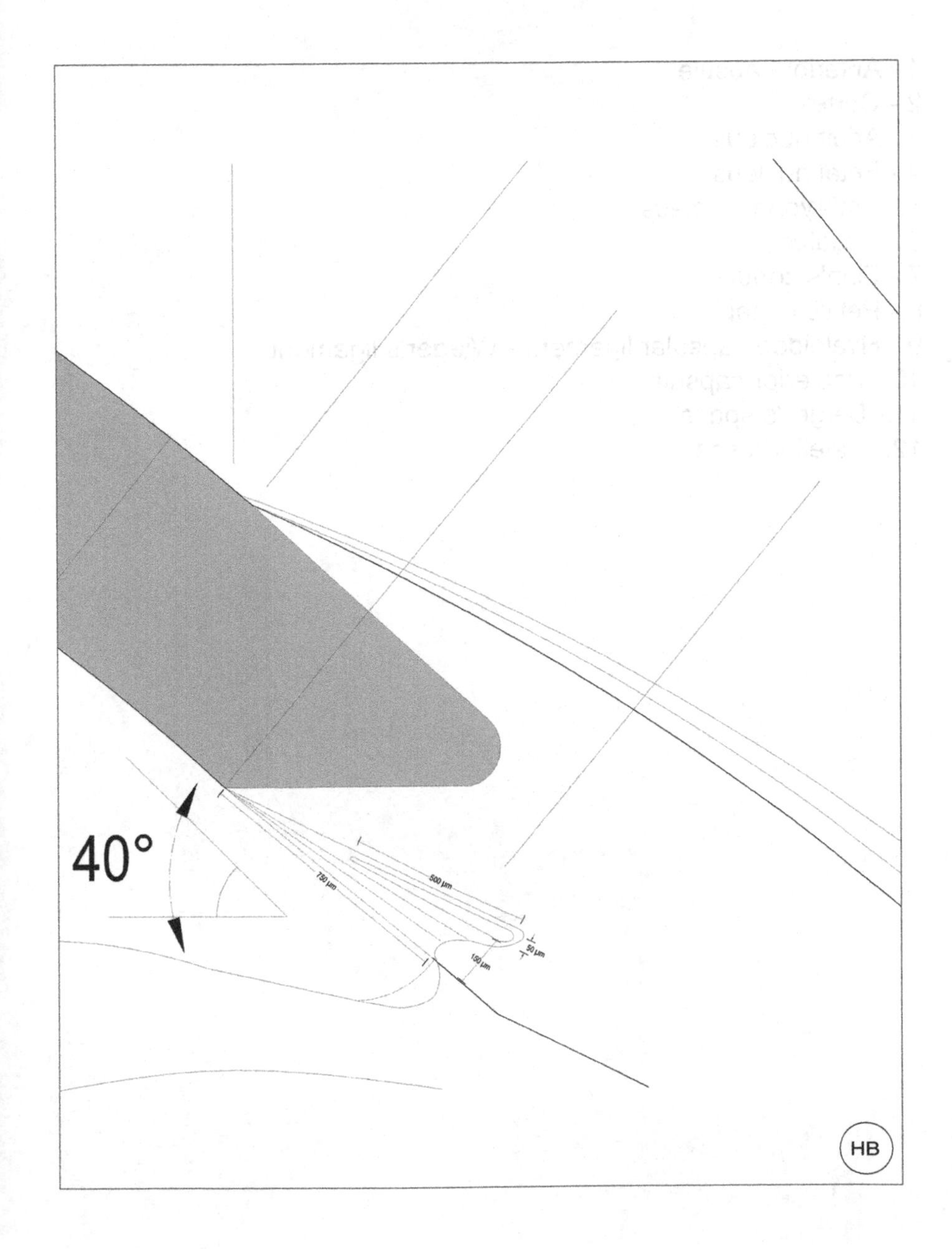

ANATOMY

Lens

1.- Anterior capsule
2.- Cortex
3.- Adult nucleus
4.- Fetal nucleus
5.- Embryonal nucleus
6.- Equator
7.- Zinn's zonule
8.- Petit's canal
9.- Hyaloideo capsular ligament - Wieger's ligament
10.- Posterior capsule
11.- Berger's space
12.- Patellar fossa

ANATOMY

Lens

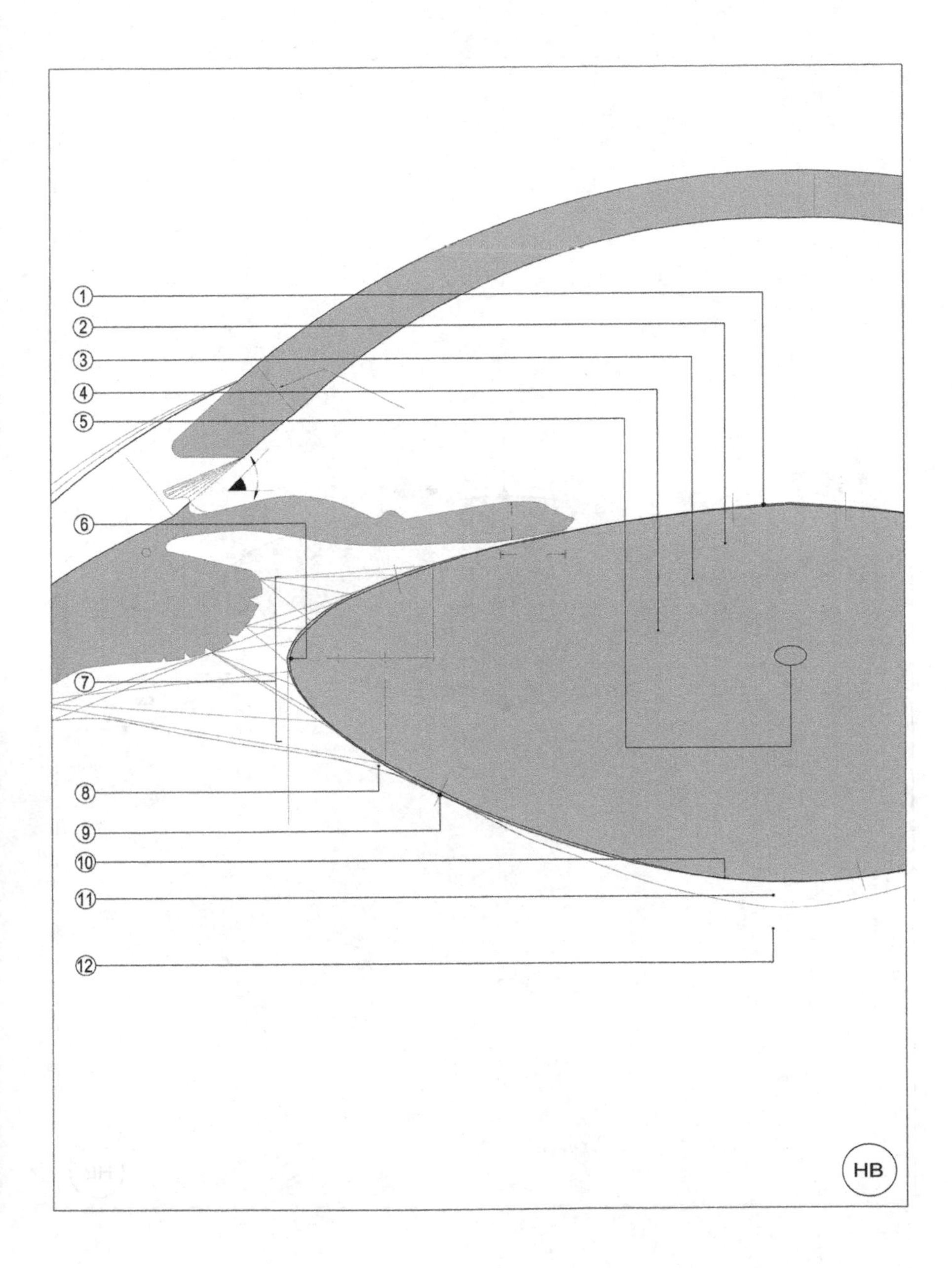

BIOMETRICS

Lens

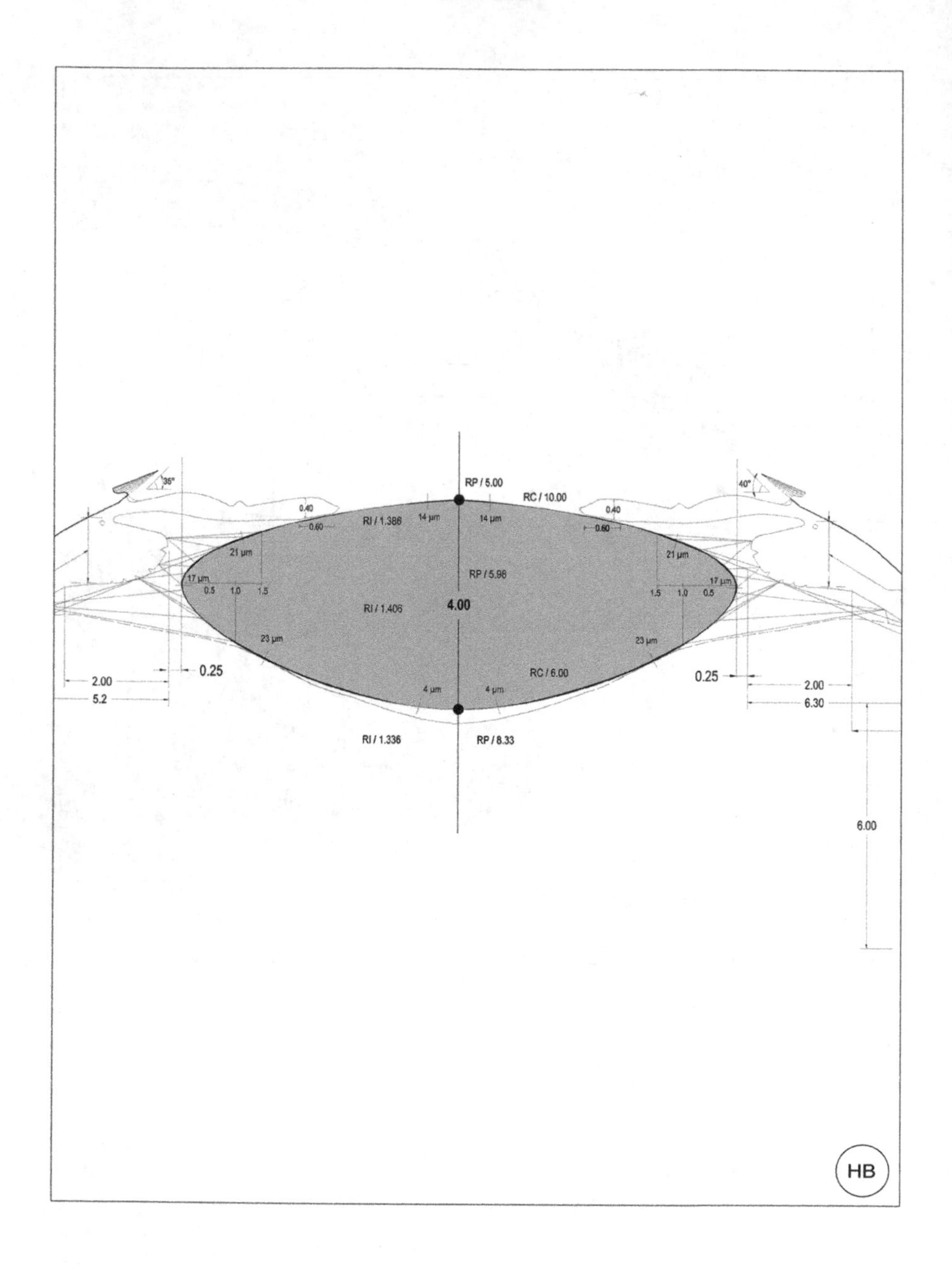

BIOMETRICS

Lens

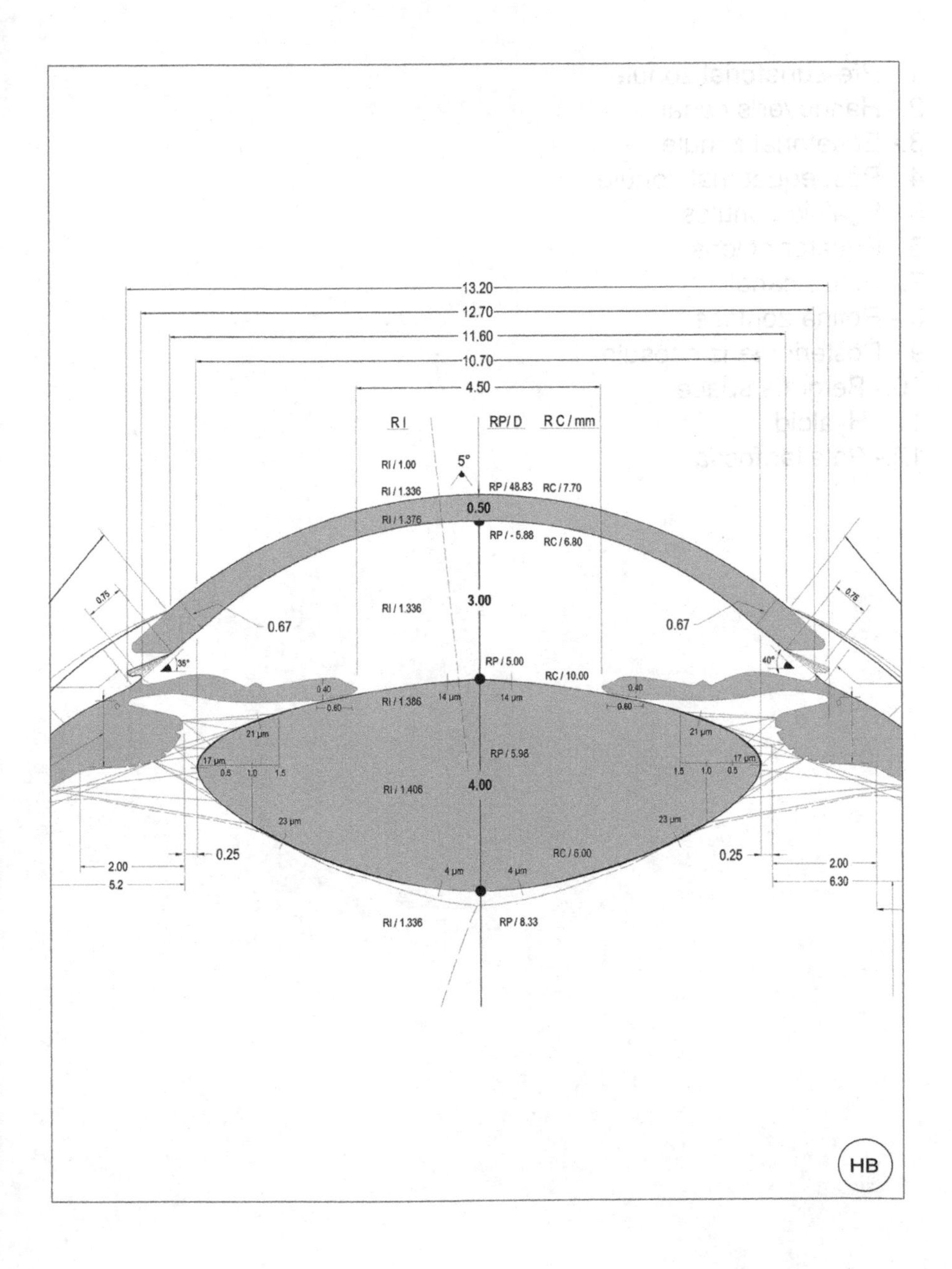

ANATOMY

Zinn's zonule

1.- Pre-equatorial zonule
2.- Hannover's canal
3.- Equatorial zonule
4.- Post-equatorial zonule
5.- Hyaloid zonules
6.- Equator of lens
7.- Petit's canal
8.- Retina zonules
9.- Posterior lens capsule
10.- Berger's space
11.- Hyaloid
12.- Patellar fossa

ANATOMY

Zinn's zonule

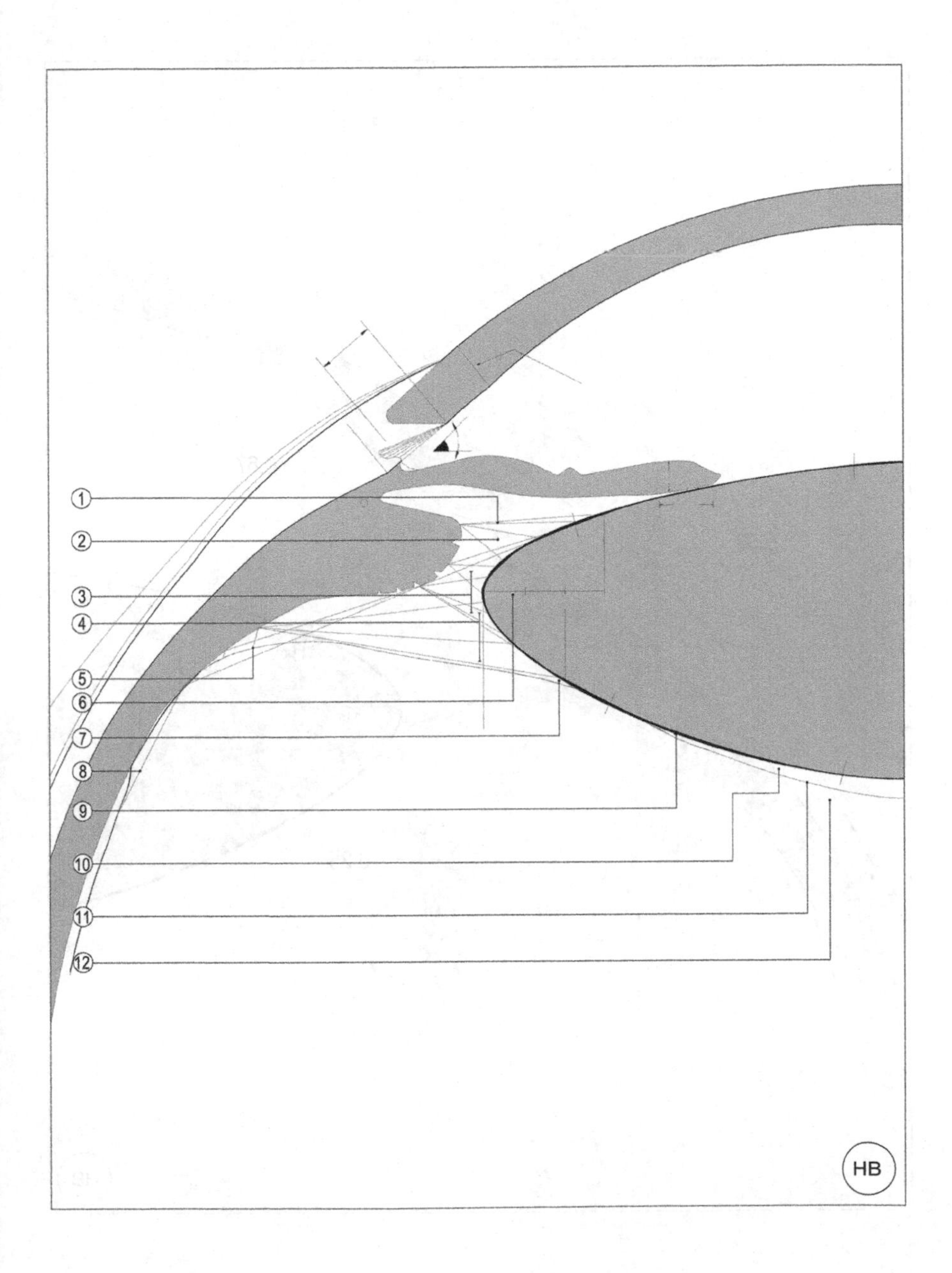

BIOMETRICS

Zinn's zonule

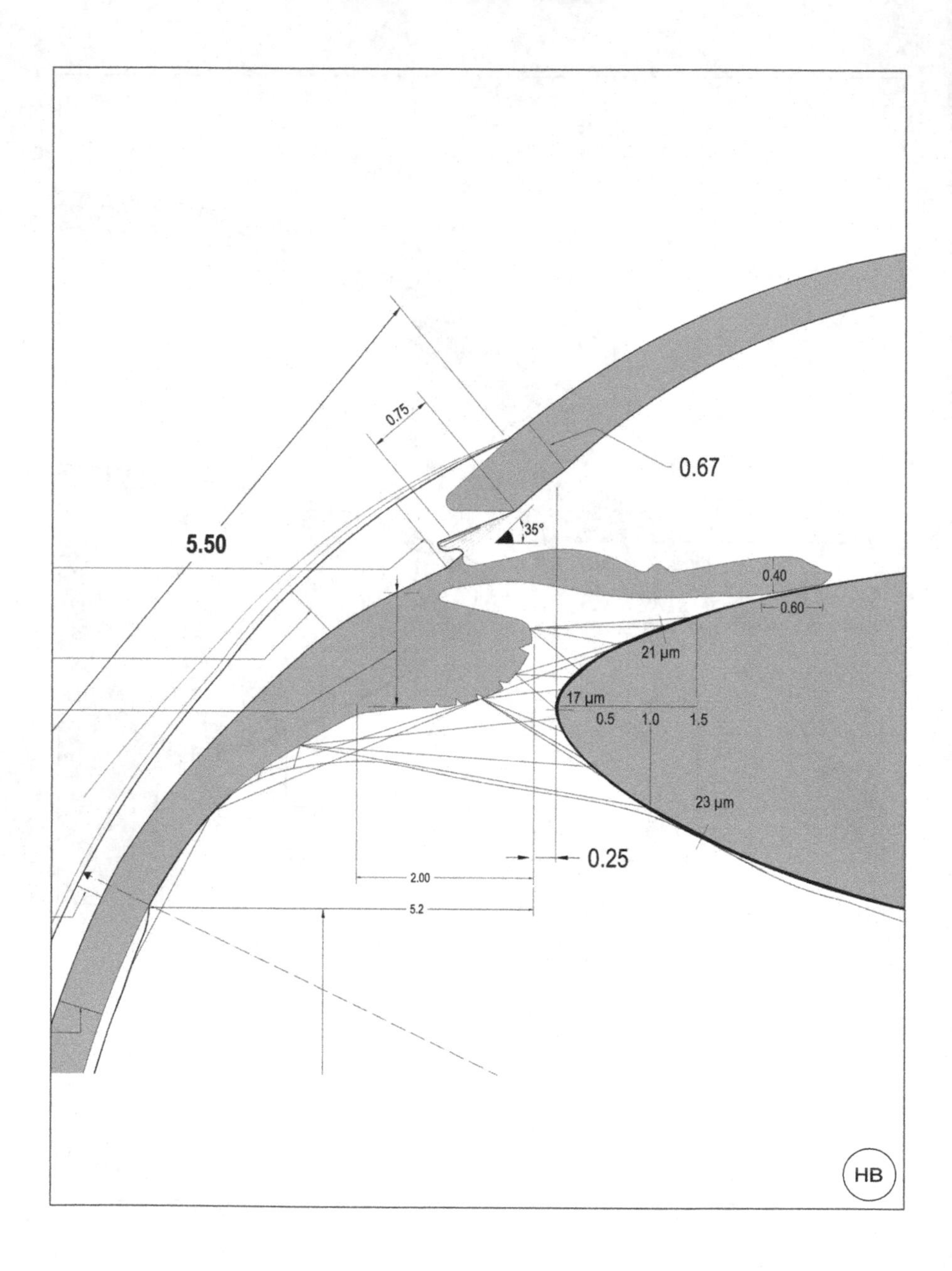

BIOMETRICS

Zinn's zonule

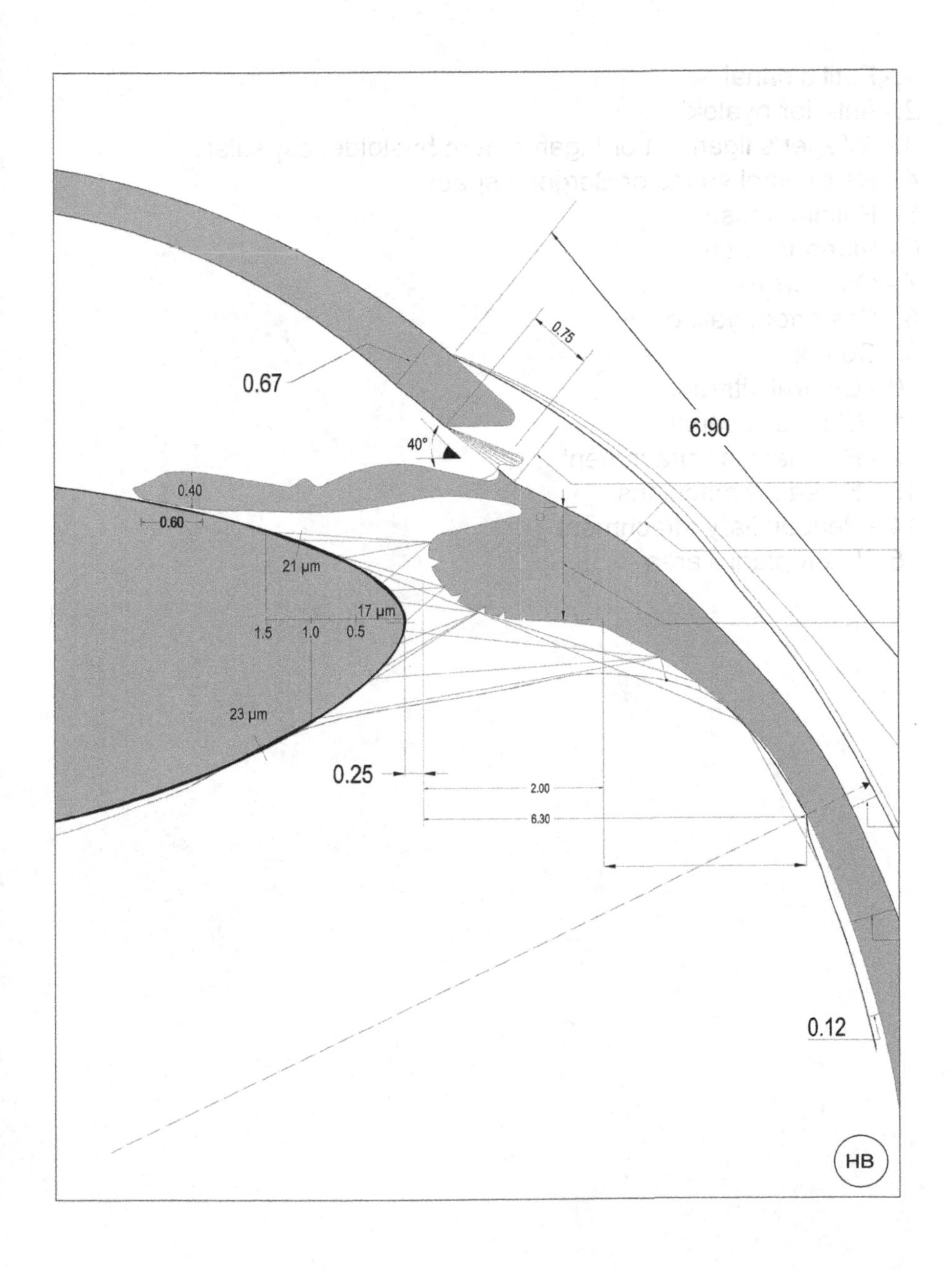

ANATOMY

Vitreous body

1.- Petit's canal
2.- Anterior hyaloid
3.- Wieger's ligament or Ligamentum hyaloideocapsular
4.- Retrolental space or Berger's space
5.- Patellar fossa
6.- Vitreous base
7.- Ora serrata
8.- Posterior hyaloid
9.- Cortex
10.- Central vitreous
11.- Cloquet's canal
12.- Perimacular attachment
13.- Bursa premacularis
14.- Peripapillary attachment
15.- Martegiani's area

ANATOMY

Vitreous body

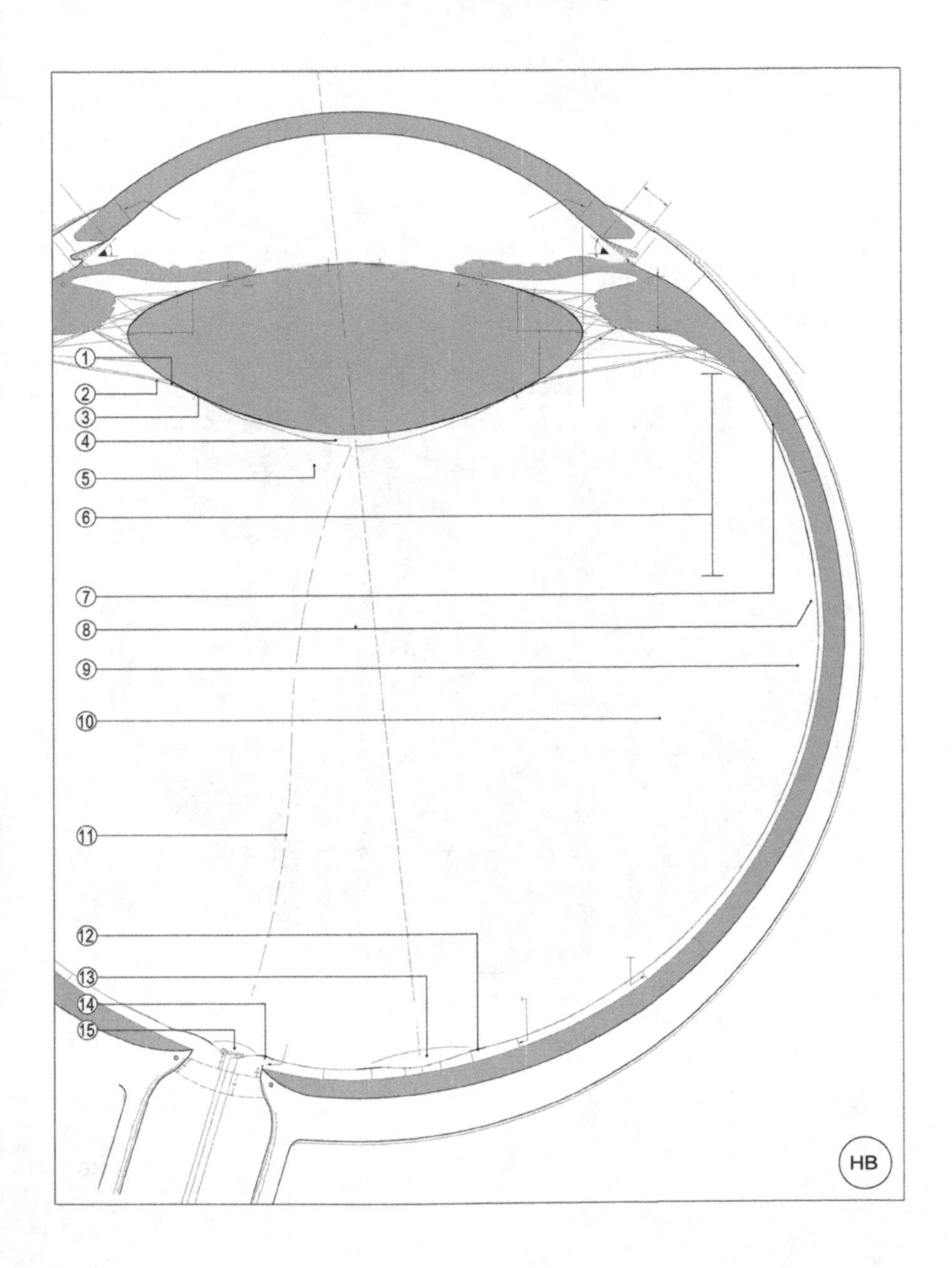

BIOMETRICS

Vitreous body

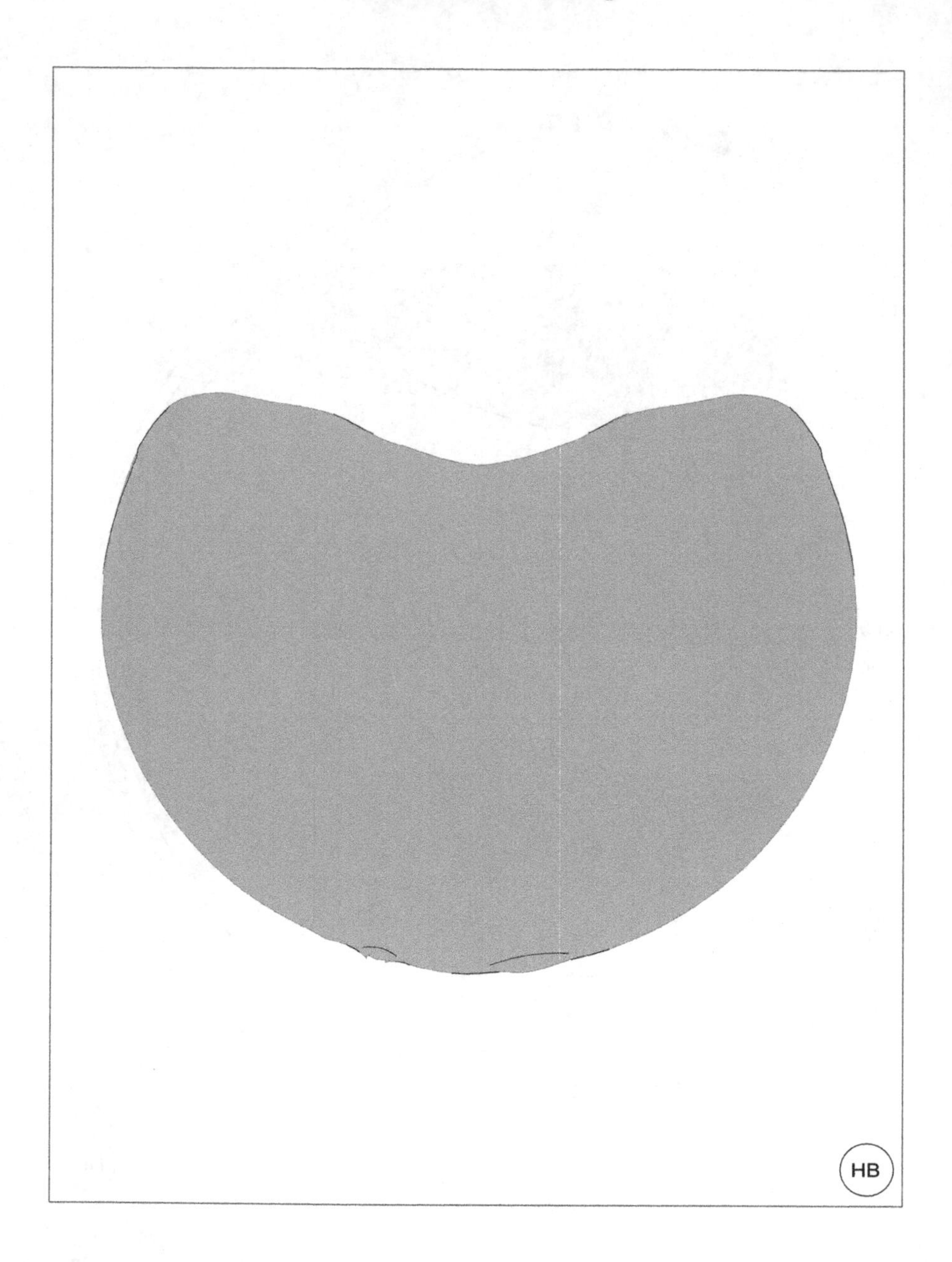

BIOMETRICS

Vitreous body

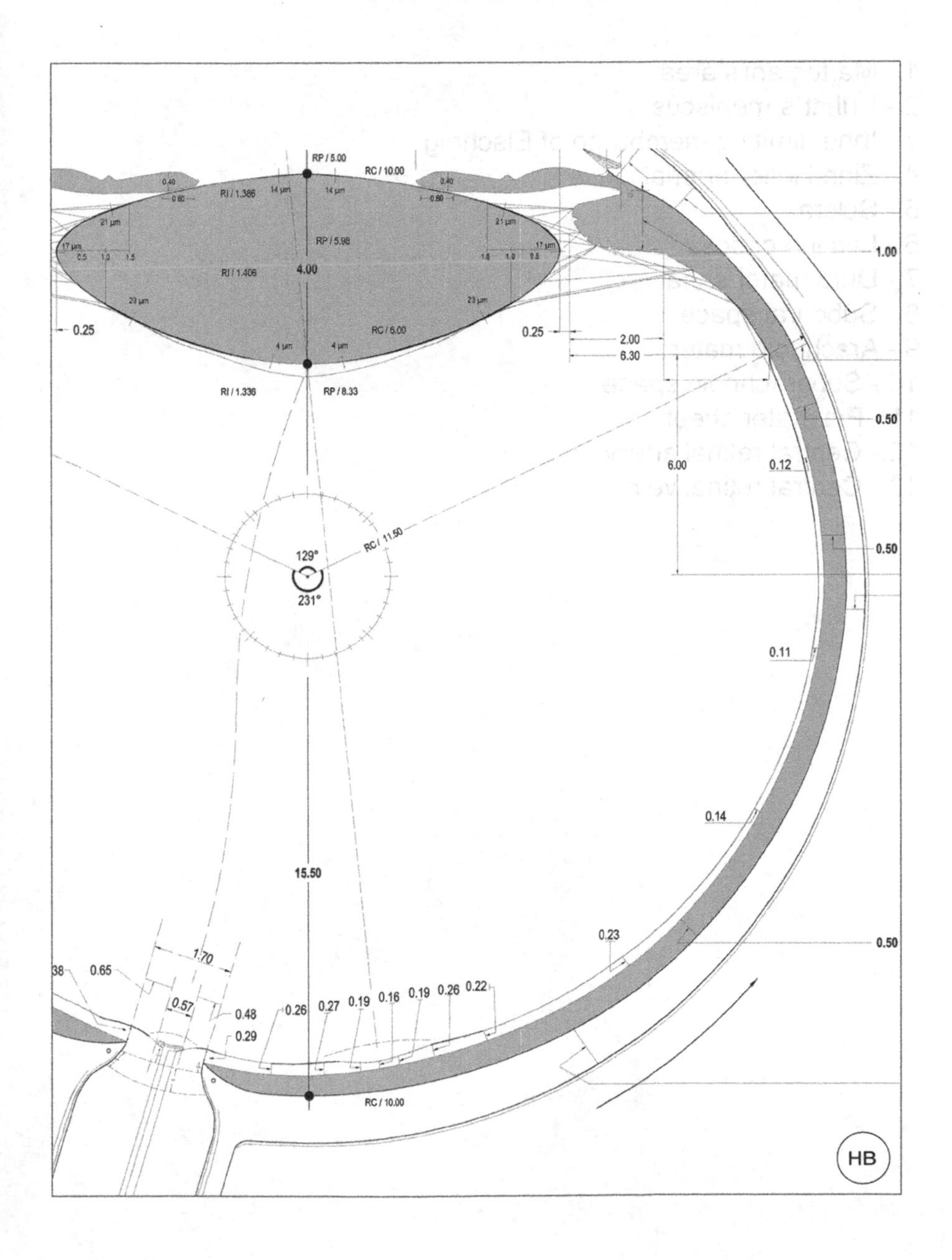

ANATOMY

Optic nerve

1.- Martegiani's área
2.- Kuhnt´s meniscus
3.- Inner limiting membrane of Elschnig
4.- Zinn-Haller arterial circle
5.- Sclera
6.- Lamina cribosa
7.- Dura mater sheat
8.- Subdural space
9.- Arachnoid mater
10.- Subarachnoid space
11.- Pia mater sheat
12.- Central retinal artery
13.- Central retinal vein

ANATOMY

Optic nerve

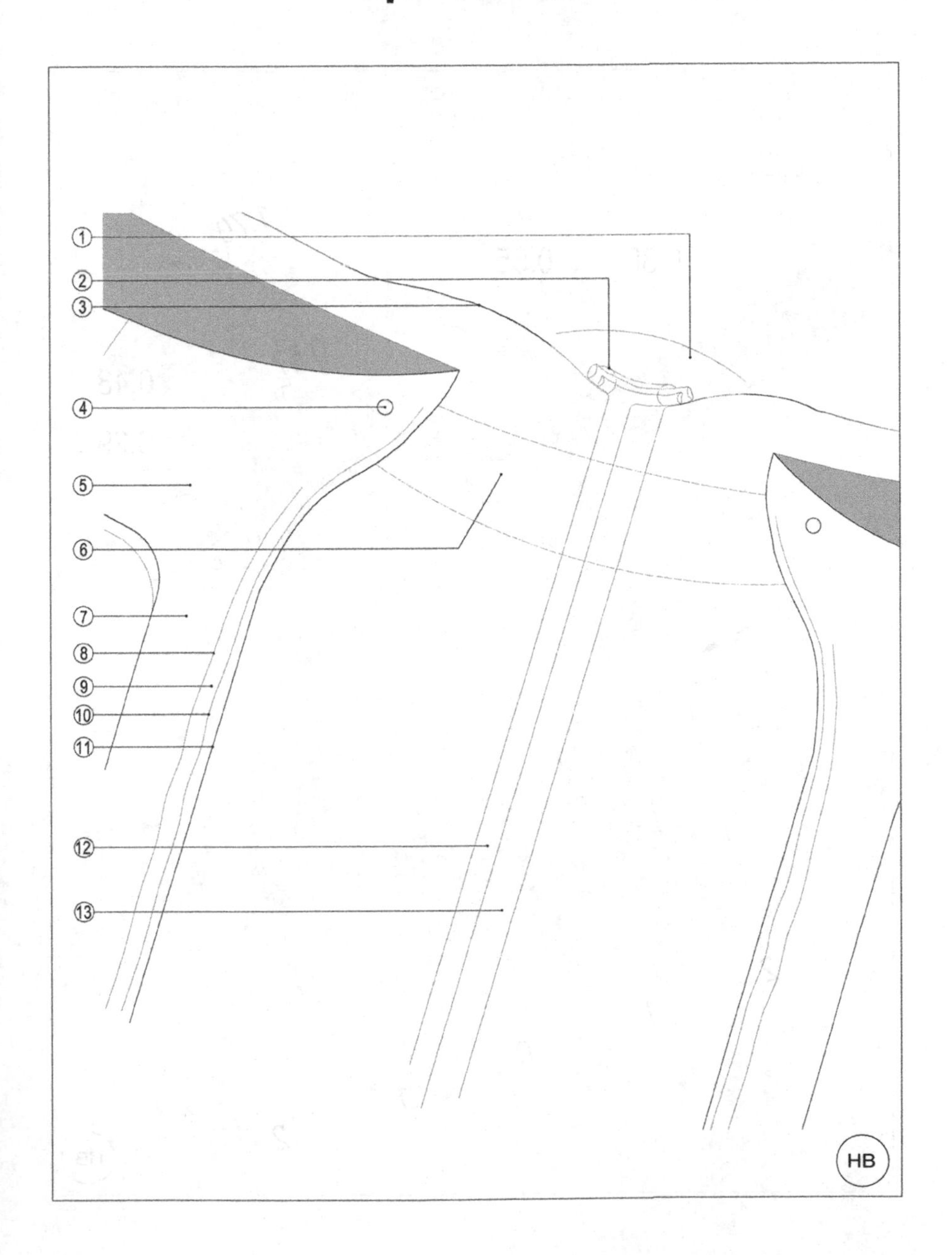

BIOMETRICS

Optic nerve

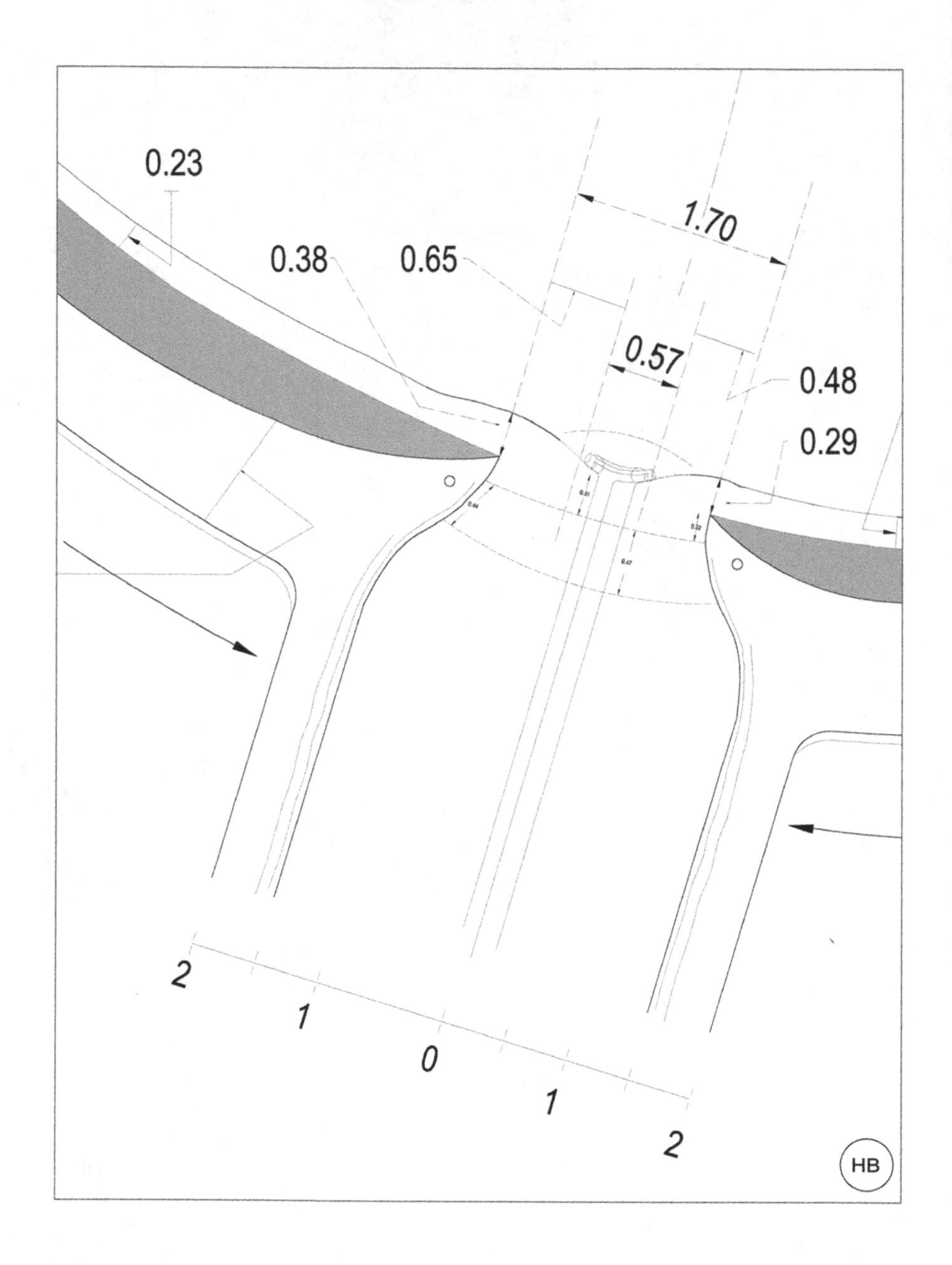

BIOMETRICS

Optic nerve

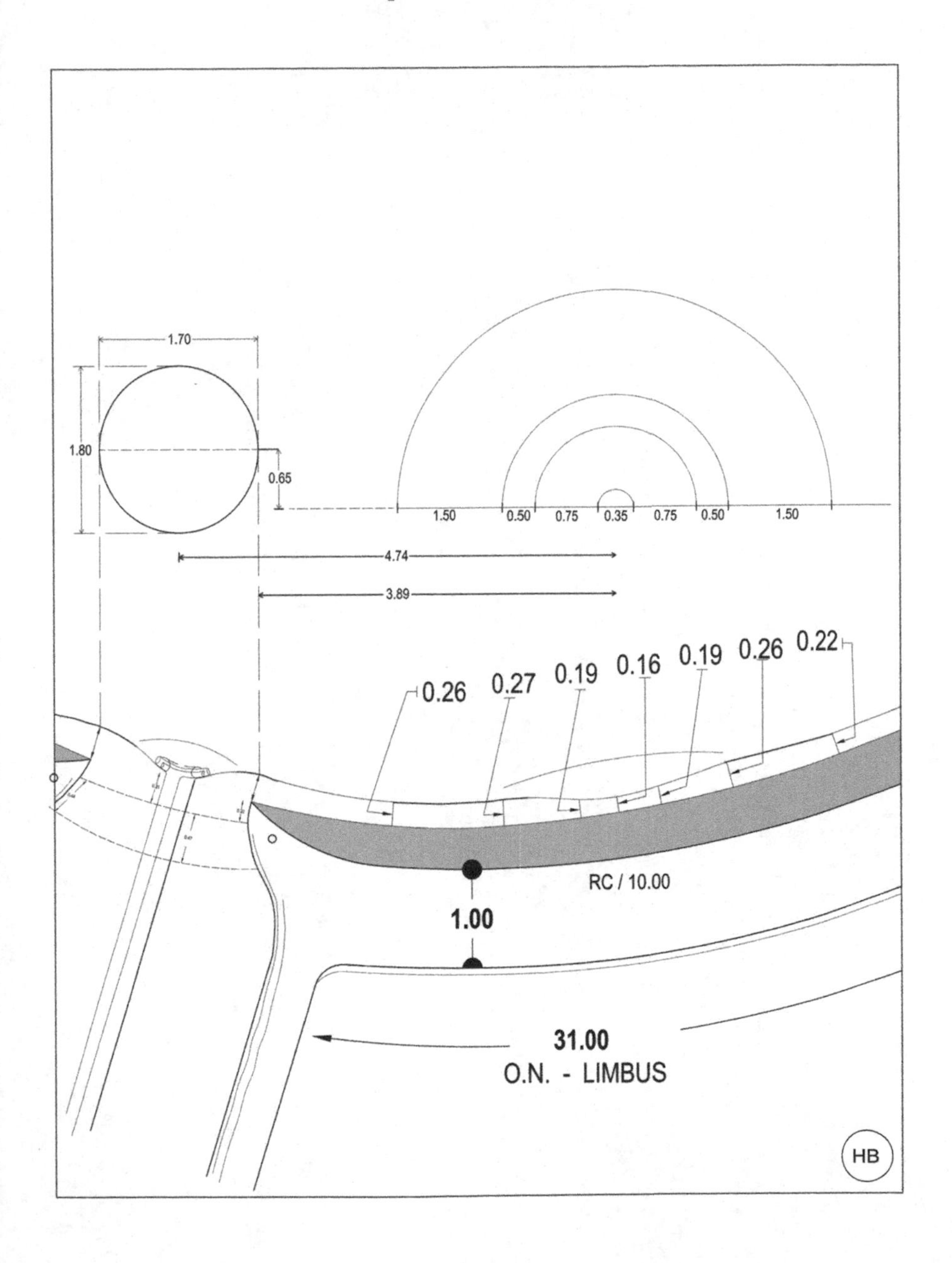

REFERENCES

1. Alkabes M, Salinas C, Vitale L, Burés-Jelstrup A, Nucci P, Mateo C. En face optical coherence tomography of inner retinal defects after internal limiting membrane peeling for idiopathic macular hole. Invest Ophthalmol Vis Sci 2011; 52: 8349–8355. | Article | PubMed |
2. Andrew J. Tatham, Atsuya Miki, Robert N. Weinreb, Linda M. Zangwill, Felipe A. Medeiros. Defects of the Lamina Cribrosa in Eyes with Localized Retinal Nerve Fiber Layer Loss. Ophthalmology, 2014;Volume 121, Issue 1, 110-118
3. Arciero, J.; Harris, A.; Siesky, B.; Amireskandari, A.; Gershuny, V.; Pickrell, A.; Guidoboni, G. Theoretical analysis of vascular regulatory mechanisms contributing to retinal blood flow autoregulation. Invest. Ophthalmol. Vis. Sci. 2013, 54, 5584–5593. [CrossRef] [PubMed]
4. Barajas H, Aranda J. Rivera A, Martinez C, Manzur I. Manzur F, Koga W,Mansilla A. The human eye anatomy.Clinical Experimental Ophthalmology 2008;36 suppl 305
5. Barak Y, Sherman MP, Schaal S. Mathematical analysis of specific anatomical foveal configurations predisposing to the formation of macular holes. Invest Ophthalmol Vis Sci 2011; 52: 8266–8270. | Article | PubMed |
6. Berdahl, J.P.; Yu, D.Y.; Morgan, W.H. The translaminar pressure gradient in sustained zero gravity, idiopathic intracranial hypertension, and glaucoma. Med. Hypotheses 2012, 79, 719–724. [Google Scholar] [CrossRef] [PubMed]
7. Berdahl, J.P.; Yu, D.Y.; Morgan, W.H. The translaminar pressure gradient in sustained zero gravity, idiopathic intracranial hypertension, and glaucoma. Med. Hypotheses 2012, 79, 719–724. [Google Scholar] [CrossRef] [PubMed]
8. Bishop PN. Structural macromolecules and supramolecular organisation of the vitreous gel. Prog Retin Eye Res 2000; 19: 323–344. | Article | PubMed |

9. Bottos J, Elizalde J, Arevalo JF, Rodrigues EB, Maia M. Vitreomacular traction syndrome. J Ophthalmic Vis Res 2012; 7: 148–161. | PubMed |
10. Bottos J, Elizalde J, Rodrigues EB, Maia M. Current concepts in vitreomacular traction syndrome. Curr Opin Ophthalmol 2012; 23: 195–201. | Article | PubMed |
11. Chang LK, Fine HF, Spaide RF, Koizumi H, Grossniklaus HE. Ultrastructural correlation of spectral-domain coherence tomographic findings in vitreomacular traction syndrome. Am J Ophthalmol 2008; 146: 121–127. | Article | PubMed |
12. Chiquet, C.; Custaud, M.A.; Le Traon, A.P.; Millet, C.; Gharib, C.; Denis, P. Changes in intraocular pressure during prolonged (7-day) head-down tilt bedrest. J. Glaucoma 2003, 12, 204–208. [Google Scholar] [CrossRef] [PubMed]
13. Christopoulos VKagemann LWollstein G et al. In vivo corneal high-speed, ultra high-resolution optical coherence tomography. Arch Ophthalmol 2007;125 (8) 1027- 1035
14. Croft MA, Nork TM, McDonald JP, Katz A, Lutjen-Drecoll E, Kaufman PL. Accommodative movements of the vitreous membrane, choroid, and sclera in young and presbyopic human and nonhuman primate eyes. Invest Ophthalmol Vis Sci. 2013;54:5049–5058.
15. Dastiridou, A.I.; Ginis, H.S.; de Brouwere, D.; Tsilimbaris, M.K.; Pallikaris, I.G. Ocular rigidity, ocular pulse amplitude, and pulsatile ocular blood flow: The effect of intraocular pressure. Investig. Ophthalmol. Vis. Sci. 2009, 50, 5718–5722. [Google Scholar] [CrossRef]
16. Delaey, C.; van de Voorde, J. Regulatory mechanisms in the retinal and choroidal circulation. Ophthalmic. Res. 2000, 32, 249–256. [Google Scholar] [CrossRef] [PubMed]
17. Drexler,Morgner, UGhanta, RKKärtner, FXSchuman, JSFujimoto JG. Ultrahigh resolution ophthalmic optical coherence tomography. Nat Med 2001;7 (4) 502- 507
18. Dua HS, Faraj LA, Said DG, Gray T, Lowe J. Human corneal anatomy redefined: a novel pre-Descemet's layer (Dua's layer). Ophthalmology. 2013;120:1778-85.

19. Elsheikh A, Geraghty B, Alhasso D, Knappett J, Campanelli M, et al. Regional variation in the biomechanical properties of the human sclera. Exp Eye Res. 2010; 90: 624–633.
20. Eric L Crowell; Mark E Gold; Alice Chuang; Laura Baker; Robert M Feldman; Nicholas P Bell; Lauren S Blieden. Characterizing Angle Landmarks with Anterior Segment Optical Coherence Tomography. Investigative Ophthalmology & Visual Science April 2014, Vol.55, 929.
21. Ezra E, Fariss RN, Possin DE, Aylward WG, Gregor ZJ, Luthert PJ et al. Immunocytochemical characterization of macular hole opercula. Arch Ophthalmol 2001; 119: 223–231. | PubMed |
22. Figueroa MS, Noval S, Contreras I. Macular structure on optical coherence tomography after lamellar macular hole surgery and its correlation with visual outcome. Can J Ophthalmol 2011; 46: 491–497. | Article | PubMed |
23. Fine HF Spaide RF. Visualization of the posterior precortical vitreous pocket in vivo with triamcinolone. Arch Ophthalmol . 2006; 124: 1663. [CrossRef] [PubMed]
24. Goldsmith JALi YChalita MR et al. Anterior chamber width measurement by high-speed optical coherence tomography. Ophthalmology 2005;112 (2) 238- 244
25. G.S. Rengifo, R. Suarez, M.O. Niño, E.L. Graue; Clinical Ophthalmologic and Genetic Study of the Corneal Dystrophies in Mexican Population . *Invest. Ophthalmol. Vis. Sci.* 2003;44(13):3870
26. Hernández-Quintela E, Mayer F, Dighiero P, Briat B, Savoldelli M, Legeais JM, Renard G. Confocal Microscopy of Cystic Disorders of the Corneal Epithelium. Ophthalmology 1998 Apr;105(4):631-6
27. Hirotaka Itakura; Shoji Kishi; Danjie Li; Hideo Akiyama.Observation of Posterior Precortical Vitreous Pocket Using Swept-Source Optical Coherence Tomography. Investigative Ophthalmology & Visual Science 2013,May Vol.54, 3102-3107
28. Hogan M, Alvarado J, Weddell J: Histology of the Human Eye—An Atlas and Textbook. Philadelphia, WB Saunders 1971
29. Hon-Tym Wong, FRCS(Ed); Marcus C. Lim, MRCS(Ed); Lisandro M. Sakata, MD, PhD; Han T. Aung, MBBS; Nishani Amerasinghe, MRCOphth; David S. Friedman, MD, MPH, PhD; Tin Aung, MBBS,

FRCS(Ed), PhD. High-Definition Optical Coherence Tomography Imaging of the Iridocorneal Angle of the Eye. .Arch Ophthalmol. 2009;127(3):256-260

30. Itakura H Kishi S. Aging changes of vitreomacular interface. Retina . 2011; 31: 1400–1404. [CrossRef] [PubMed]
31. Itakura H, Kishi S, Li D, Akiyama H. Observation of posterior precortical vitreous pocket using swept-source optical coherence tomography. Invest Ophthalmol Vis Sci 2013; 54: 3102–3107. | Article | PubMed |
32. Jackson TL, Nicod E, Simpson A, Angelis A, Grimaccia F, Kanavos P. Symptomatic vitreomacular adhesion. Retina 2013; 33: 1503–1511. | Article | PubMed |
33. Johnson GJ. The environment and the eye. Eye. 2004;18:1235-50.
34. Johnson MW. Perifoveal vitreous detachment and its macular complications. Trans Am Ophthalmol Soc . 2005; 103: 537–567. [PubMed]
35. Johnson MW. Posterior vitreous detachment: evolution and complications of its early stages. Am J Ophthalmol 2010; 149: 371–382. | Article | PubMed |
36. Jonas, J.B. Trans-lamina cribrosa pressure difference. Arch. Ophthalmol. 2007, 125, 431–431. [Google Scholar] [CrossRef] [PubMed]
37. Jonas J.B. Optic disk size correlated with refractive error. Am J Ophthalmol. 2005; 139: 346–348.
38. Jonas, J.B.; Nangia, V.; Matin, A.; Sinha, A.; Kulkarni, M.; Bhojwani, K. Intraocular pressure and associated factors: The Central India Eye and Medical Study. J. Glaucoma 2011, 20, 405–409. [Google Scholar] [CrossRef] [PubMed]
39. Kampik A. Pathology of epiretinal membrane, idiopathic macular hole, and vitreomacular traction syndrome. Retina 2012; 32: 194–199. | Article | PubMed |
40. Katrin Petermeier, Daniela Suesskind, Elke Altpeter, Andreas Schatz, André Messias, Florian Gekeler, Peter Szurman. Sulcus anatomy and diameter in pseudophakic eyes and correlation with biometric data: Evaluation with a 50 MHz ultrasound

biomicroscope.Journal of Cataract & Refractive Surgery, 2012; Volume 38, Issue 6, 986-991

41. Kiernan DF, Mieler WF, Hariprasad SM. Spectral-domain optical coherence tomography: a comparison of modern high-resolution retinal imaging systems. Am J Ophthalmol 2010; 149: 18–31. | Article | PubMed
42. Kishi S Hagimura N Shimizu K. The role of the premacular liquefied pocket and premacular vitreous cortex in idiopathic macular hole development. Am J Ophthalmol . 1996; 122: 622–628. [CrossRef] [PubMed]
43. Knight OJ, Girkin CA, Budenz DL, Durbin MK, Feuer WJ; Cirrus OCT Normative Database Study Group. Effect of race, age, and axial length on optic nerve head parameters and retinal nerve fiber layer thickness measured by Cirrus HD-OCT. Arch Ophthalmol. 2012 Mar;130(3):312-8.
44. Krebs, I., Brannath, W., Glittenberg, K., et al. Posterior vitreomacular adhesion: A potential risk factor for exudative age-related macular degeneration. American Journal of Ophthalmology. 2007; 144: 741–746.
45. Kumagai K, Ogino N, Hangai M, Larson E. Percentage of fellow eyes that develop full-thickness macular hole in patients with unilateral macular hole. Arch Ophthalmol 2012; 130: 393–394. | Article | PubMed
46. Kur, J.; Newman, E.A.; Chan-Ling, T. Cellular and physiological mechanisms underlying blood flow regulation in the retina and choroid in health and disease. Prog. Retin. Eye Res. 2012, 31, 377–406. [Google Scholar] [CrossRef] [PubMed]
47. Levin L, Nilsson S, Ver Hoeve J, Wu S, Kaufman P, Alm A. Adler's physiology of the eye. 11ª ed. Edinburgh: Elsevier Inc.; 2011
48. Longo, A.; Geiser, M.H.; Riva, C.E. Posture changes and subfoveal choroidal blood flow. Invest. Ophthalmol. Vis. Sci. 2004, 45, 546–551. [Google Scholar] [CrossRef] [PubMed]
49. Mader, T.H.; Gibson, C.R.; Pass, A.F.; Kramer, L.A.; Lee, A.G.; Fogarty, J.; Tarver, W.J.; Dervay, J.P.; Hamilton, D.R.; Sargsyan, A.; et al. Optic disc edema, globe glattening, choroidal folds, and hyperopic shifts observed in astronauts after long-duration space

flight. Ophthalmology 2011, 118, 2058–2069. [Google Scholar] [CrossRef] [PubMed]

50. MansillaA. BarajasH.,etal. Theoreticalaspectsoftheneurobiological integration of memory. Medical Hypotheses.2000,54(1),51-58

51. Michalewska Z, Michalewski J, Odrobina D, Nawrocki J. Non-full thickness macular holes reassessed with spectral domain optical coherence tomography. Retina 2012; 32: 922–929. | Article | PubMed |

52. Miura M, Elsner AE, Osako M, Iwasaki T, Okano T, Usui M. Dissociated optic nerve fiber layer appearance after internal limiting membrane peeling for idiopathic macular hole. Retina 2003; 23: 561–563. | Article | PubMed |

53. Mohamed-Noor J, Bochmann F, Siddiqui MA, Atta HR, Leslie T, et al. Correlation between corneal and scleral thickness in glaucoma. J Glaucoma. 2009; 18: 32–36.

54. Mojana F Kozak I Oster SF Observations by spectral-domain optical coherence tomography combined with simultaneous scanning laser ophthalmoscopy: imaging of the vitreous. Am J Ophthalmol . 2010; 149: 641–650. [CrossRef] [PubMed]

55. Morgan, A.J.; Hosking, S.L. Non-invasive vascular impedance measures demonstrate ocular vasoconstriction during isometric exercise. Br. J. Ophthalmol. 2007, 91, 385–390. [Google Scholar] [CrossRef]

56. Mühlendyck H: Wachstum und Lange der ausseren Augenmuskeln. Ber Dtsch Ophthalmol Ges 75:449, 1978.

57. Mwanza J-C, Chang RT, Budenz DL, Durbin MK, Gendy MG, Shi W, et al. Reproducibility of Peripapillary Retinal Nerve Fiber Layer Thickness and Optic Nerve Head Parameters Measured with Cirrus HD-OCT in Glaucomatous Eyes. Invest Ophthalmol Vis Sci. 2010 Nov;51(11):5724-30.

58. Nagahara, M.; Tamaki, Y.; Tomidokoro, A.; Araie, M. In vivo measurement of blood velocity in human major retinal vessels using the laser speckle method. Investig. Ophthalmol. Vis. Sci. 2011, 52, 87–92. [Google Scholar] [CrossRef]

59. Nickla, D.L.; Wallman, J. The multifunctional choroid. Progr. Retin. Eye Res. 2010, 29, 144–168. [Google Scholar] [CrossRef]

60. Norman RE, Flanagan JG, Rausch SM, Sigal IA, Tertinegg I, et al. Dimensions of the human sclera: Thickness measurement and regional changes with axial length. Exp Eye Res. 2010; 90: 277–84.
61. Norman RE, Flanagan JG, Sigal IA, Rausch SM, Tertinegg I, et al. Finite element modeling of the human sclera: Influence on optic nerve head biomechanics and connections with glaucoma. Exp Eye Res 2011; 93: 4–12.
62. Odrobina D, Michalewska Z, Michalewski J, Dziegielewski K, Nawrocki J. Long-term evaluation of vitreomacular traction disorder in spectral-domain optical coherence tomography. Retina 2011; 31: 324–331. | Article | PubMed |
63. Ohno-Matsui K, Kasahara K, Moriyama M. Detection of Zinn-Haller arterial ring in highly myopic eyes by simultaneous indocyanine green angiography and optical coherence tomography. Am J Ophthalmol. 2013;155:920-6.
64. Pavlin CJ, Sherar MD, Foster FS. Subsurface ultrasound microscopic imaging of the intact eye. Ophthalmology 1990; 97(2): 244–250. | PubMed |
65. Pavlin CJ, McWhae JA, McGowan HD, Foster FS. Ultrasound biomicroscopy of anterior segment tumors. Ophthalmology 1992; 99(8): 1220–1228. | PubMed | ISI | ChemPort |
66. Pavlin CJ, Harasiewicz K, Foster FS. Ultrasound biomicroscopy of anterior segment structures in normal and glaucomatous eyes. Am J Ophthalmol 1992; 113(4): 381–389. | PubMed | ISI | ChemPort |
67. Pavlin CJ. Ultrasound Biomicroscopy of the Eye. New York,N.Y: Springer-Verlag;1995
68. Pavlin CJ. Ciliary Body Detachment. Ophthalmology. October 2011Volume 118, Issue 10, Pages 2097–2097.
69. Pavlin CJ. Ultrasound biomicroscopy in glaucoma.Acta Ophthalmologica. 1992; 70, 204, 7–9
70. Radhakrishnan SGoldsmith JHuang D et al. Comparison of optical coherence tomography and ultrasound biomicroscopy for detection of narrow anterior chamber angles. Arch Ophthalmol 2005;123 (8) 1053- 1059

71. Radhakrishnan SRollins AMRoth JE et al. Real-time optical coherence tomography of the anterior segment at 1310 nm. Arch Ophthalmol 2001;119 (8) 1179- 1185
72. Reibaldi M, Parravano M, Varano M, Longo A, Avitabile T, Uva MG et al. Foveal microstructure and functional parameters in lamellar macular hole. Am J Ophthalmol 2012; 154: 974–980. | Article | PubMed |
73. Robison, C., Krebs, I., Binder, S., et al. (2009). Vitreo-macular adhesion in active and end-stage age-related macular degeneration. American Journal of Ophthalmology 148(1): 79–82.
74. Ruiz-Moreno JM, Staicu C, Piñero DP, Montero J, Lugo F, Amat P. Optical coherence tomography predictive factors for macular hole surgery outcome. Br J Ophthalmol 2008; 92: 640–644. | Article | PubMed |
75. Sanchez Huerta V, De Wit Carter G. Hernández-Quintela E, Naranjo-Tackman R. Occupational corneal argyrosis in art silver solderers. Cornea 2003 Oct 22:7 604-11
76. Shinojima, A.; Iwasaki, K.; Aoki, K.; Ogawa, Y.; Yanagida, R.; Yuzawa, M. Subfoveal choroidal thickness and foveal retinal thickness during head-down tilt. Aviat. Space Environ. Med. 2012, 83, 388–393. [Google Scholar] [CrossRef] [PubMed]
77. Sigal IA, Bilonick RA, Kagemann L, Wollstein G, Ishikawa H, Schuman JS, et al. The optic nerve head as a robust biomechanical system. Invest Ophthalmol Vis Sci. 2012 May;53(6):2658-67.
78. Sigal IA, Flanagan JG, Ethier CR . Factors influencing optic nerve head biomechanics. Invest Ophthalmol Vis Sci. 2005; 46: 4189–4199. |
79. Sigal IA, Flanagan JG, Lathrop KL, Tertinegg I, Bilonick R. Human lamina cribrosa insertion and age. Invest Ophthalmol Vis Sci. 2012 Ago;53(11):6870-9.
80. Sigal IA, Flanagan JG, Tertinegg I, Ethier CR. 3D morphometry of the human optic nerve head. Exp Eye Res. 2010; 90: 70–8
81. Sigal IA, Flanagan JG, Tertinegg I, Ethier CR. Finite Element Modeling of Optic Nerve Head Biomechanics. Invest Ophthalmol Vis Sci. 2004 Ago;45 (12):4378-87.

82. Silver, D.M.; Geyer, O. Pressure-volume relation for the living human eye. Curr. Eye Res. 2000, 20, 115–120. [Google Scholar] [CrossRef] [PubMed]
83. Stalmans, I.; Vandewalle, E.; Anderson, D.R.; Costa, V.P.; Frenkel, R.E.; Garhofer, G.; Grunwald, J.; Gugleta, K.; Harris, A.; Hudson, C.; et al. Use of colour Doppler imaging in ocular blood flow research. Acta Ophthalmol. 2011, 89, 609–630. [Google Scholar] [CrossRef] [PubMed]
84. Taibbi, G.; Cromwell, R.L.; Kapoor, K.G.; Godley, B.F.; Vizzeri, G. The effect of microgravity on ocular structures and visual function: A review. Surv. Ophthalmol. 2013, 58, 155–163. [Google Scholar] [CrossRef] [PubMed]
85. Thornton, I.L.; Dupps, W.J.; Roy, A.S.; Krueger, R.R. Biomechanical effects of intraocular pressure elevation on optic nerve/lamina cribrosa before and after peripapillary scleral collagen cross-linking. Investig. Ophthalmol. Vis. Sci. 2009, 50, 1227–1233. [Google Scholar] [CrossRef]
86. Uchino E Uemura A Ohba N. Initial stages of posterior vitreous detachment in healthy eyes of older persons evaluated by optical coherence tomography. Arch Ophthalmol . 2001; 119: 1475–1479. [CrossRef] [PubMed]
87. Velez-Montoya R, Guerrero-Naranjo JL, Gonzalez-Mijares CC, Fromow-Guerra J, Marcellino GR, Quiroz-Mercado H, Morales-Cantón V. Pattern scan laser photocoagulation: safety and complications, experience after 1301 consecutive cases. Br J Ophthalmol. 2010 Jun;94(6):720-4.
88. Wang S, Xu L, Jonas JB. Prevalence of full-thickness macular holes in urban and rural adult Chinese: the Beijing Eye Study. Am J Ophthalmol 2006; 141: 589–591. | Article | PubMed |
89. Wang, Y.; Lu, A.; Gil-Flamer, J.; Tan, O.; Izatt, J.A.; Huang, D. Measurement of total blood flow in the normal human retina using Doppler Fourier-domain optical coherence tomography. Br. J. Ophthalmol. 2009, 93, 634–637. [Google Scholar] [CrossRef] [PubMed]
90. Werkmeister, R.M.; Dragostinoff, N.; Palkovits, S.; Told, R.; Boltz, A.; Leitgeb, R.A.; Groschl, M.; Garhofer, G.; Schmetterer, L.

Measurement of absolute blood flow velocity and blood flow in the human retina by dual-beam bidirectional Doppler fourier-domain optical coherence tomography. Invest. Ophthalmol. Vis. Sci. 2012, 53, 6062–6071. [Google Scholar] [CrossRef] [PubMed]

91. Wilson SE, Hong JW. Bowman's layer structure and function: critical or dispensable to corneal function? A hypothesis. Cornea. 2000;4: 417-22.

92. Wollensak G, Spoerl E, Grosse G, Wirbelauer C. Biomechanical significance of the human internal limiting lamina. Retina 2006; 26: 965–968. | Article | PubMed |

93. Xu D, Yuan A, Kaiser PK, Srivastava SK, Singh RP, Sears JE et al. A novel segmentation algorithm for volumetric analysis of macular hole boundaries identified with optical coherence tomography. Invest Ophthalmol Vis Sci 2013; 54: 163–169. | Article | PubMed |

94. Xu I. Anterior chamber depth and chamber angle. Am J. Ophthal. 2008;145: 929-936

95. Xu L, Wang YX, Wang S, Jonas JB. Definition of high myopia by parapapillary atrophy. The Beijing Eye Study. Acta Ophthalmol. 2010; 88: 350–351.

96. Yasuda A, Yamaguchi T, Ohkoshi K. Changes in corneal curvature in accommodation. J Catarct Refract Surg. 2003;29:1297-301.

INDEX

Printed in the United States
By Bookmasters